THE PHYSICS OF MAGNETIC RECORDING

by

C.D. MEE

IBM Research Laboratory,
formerly with CBS Laboratories.

NORTH-HOLLAND
AMSTERDAM • OXFORD • NEW YORK • TOKYO

ISBN: 0 444 87043 1

Hardbound edition 1964
Reprinted 1968
   (Series: Selected Topics in Solid State Physics)
Paperback edition 1986
   (Series: North-Holland Personal Library)

Published by:
North-Holland Physics Publishing
a division of
Elsevier Science Publishers B.V.
P.O. Box 103
1000 AC Amsterdam
The Netherlands

$$D$$
$$621.3893'2$$
$$MEE$$

Sole distributors for the USA and Canada:
Elsevier Science Publishing Company, Inc
52 Vanderbilt Avenue
New York, NY 10017
USA

Library of Congress Catalog Card Number: 86-31274

Printed in the Netherlands

Dedicated to Professor L.F. Bates, F.R.S.

PUBLISHER'S PREFACE

Twenty-two years have passed since the first printing of this book. Many important new developments occurred in the meantime in magnetic recording technologies. Tape recording, the main subject of the book, is still the most important method, although others have emerged and have seen rapid development. While new materials found application in magnetic tapes, the principles magnetic recording and reproducing systems are based on remain unchanged. This partly explains why the excellent presentation of the "state of the art" in 1964 by Dr. C.D. Mee is still found valuable by those active in developing new materials and techniques in magnetic recording.

The response to the first two printings, as volume 2 in the series of monographs on Selected Topics in Solid State Physics edited by E.P. Wohlfarth, and the continued demand for the book clearly indicate that this third printing, published in the North-Holland Personal Library series, will be appreciated by the community. While the text is gradually acquiring the status of a classic, it is by no means of solely historical interest and it will serve new generations of scientists as a valuable introduction and a standard reference.

PREFACE

Magnetic tape recording is an important modern technology, and a large number of engineers, physicists and chemists throughout the world currently devote their efforts towards its further development. Despite the ever increasing number of applications for magnetic tape recording and the continuing increase of published literature on this subject, the basic techniques and components in use today are very similar to those developed for military use during the Second World War. The aim of this book is to attempt to describe these techniques and components from the point of view of the basic magnetization mechanisms involved; the scope and contents are described in the introductory chapter.

I have attempted to include full reference to other writers, whose work I have quoted, and who have kindly given permission for inclusion of their diagrams and photographs. Inevitably, some omissions can occur in this respect, and, also, some important publications in the field of magnetic recording will have been missed; I hope that such omissions are few. Future important work can certainly be expected within the scope of this book and updating of the text is anticipated.

I would like to record my appreciation to colleagues and friends who have contributed directly and indirectly towards the production of this book. I am especially grateful to my former associates at MSS Recording Company, England, and CBS Laboratories, U.S.A. from whom I have learned much about magnetic recording. I am indebted to Dr. E. P. Wohlfarth for inviting me to write this book and for his guidance and counsel on its contents. Valuable assistance with the organization of the book has been given by Mrs. M. E. C. Mee through all stages of its preparation. Messrs. E. D. Daniel, J. C. Jeschke, G. Y. Fan and G. Bate have also kindly helped by critically reading various chapters.

C. D. Mee

CONTENTS

LIST OF MOST IMPORTANT SYMBOLS

I_0 — Intensity of magnetization for applied dc field small compared to H_c

I_s — Saturation intensity of magnetization

$I(H_{dc})$ — Intensity of magnetization for constant applied field H_{dc}

$I_r(H_{dc})$ — Remanent intensity of magnetization from constant applied field H_{dc} (maximum value I_{rs})

$I_r(t)$ — Remanent intensity of magnetization at time t

$I_{ar}''(H_{dc})$ — Anhysteretic remanent intensity of magnetization from constant applied field H_{dc}

$I_{ar}'(H_{dc})$ — Anhysteretic remanent intensity of magnetization from constant applied field H_{dc} acquired in presence of demagnetizing field

$I_{ar}''(H_{dc} \to 0)$ — Anhysteretic remanent intensity of magnetization from initial applied field H_{dc} falling to zero with $H_{dc}/H_{ac} =$ constant

$I(H_i, H_c')$ — Intensity of magnetization of element with switching field H_c' and internal field H_i

$I_d''(H_{dc})$ — Remanent magnetization after dc demagnetization

$I_d''(H_{ac})$ — Remanent magnetization after ac demagnetization

I_s'' — Tape saturation intensity of magnetization

I_r'' — Tape remanent intensity of magnetization

$I_r''(H_{sig})$ — Tape remanent intensity of magnetization for recording head signal field H_{sig}

$I_{ar}''(H_{sig})$ — Tape anhysteretic remanent intensity of magnetization for initial signal field, H_{sig}, falling to zero with $H_{sig}/H_{ac} =$ constant

j_r — Reduced remanent intensity of magnetization

σ_s — Specific saturation magnetization at room temperature

σ_{s0} — Specific saturation magnetization at $0°$ K

σ	Specific magnetization, or intensity of magnetization/density of material
m	Particle magnetic moment
χ_d''	Tape differential susceptibility
χ_r''	Tape reversible susceptibility
η_t''	Tape sensitivity ($\eta_t'' = I_{ar}''(H_{sig})/H_x$)
s	Initial anhysteretic susceptibility
s'	Initial anhysteretic susceptibility in presence of demagnetizing field
s_μ	Initial anhysteretic susceptibility with $H_{dc}/H_{ac} =$ constant
N	Demagnetizing coefficient
N_e	External demagnetizing coefficient ($H_d = N_e I$)
N_i	Internal demagnetizing coefficient
B	Magnetic induction
B_i	Intrinsic induction ($= B-H$)
B_{ir}	Remanent intrinsic induction ($= B_r$)
B_y''	Tape normal surface induction
μ_r''	Tape reversible permeability
φ_i	Intrinsic flux
η_r''	Tape recording sensitivity $= \dfrac{\text{tape flux}}{\text{field in head gap}}$
H	Applied magnetic field
H_s	Applied magnetic field producing saturation of irreversible magnetization
H_{dc}	Constant applied magnetic field
H_{ac}	High frequency alternating magnetic field
H_{sig}	Signal magnetic field
H_c	Intrinsic coercive field
H_c'	Intrinsic critical field for irreversible switching of single particle
H_d	External demagnetizing field
H_i	Internal field
H_{xw}	Recording head magnetic field (longitudinal component)
H_{xr}	Reproducing head magnetic field (longitudinal component)

H_{dc_0}	Dc field in head gap
H_{ac_0}	Ac field in head gap
H_a	Anisotropy field
H_{at}	Anisotropy field at 150°C
i_s	Recording head signal current
i_b	Recording head bias current
g	Recording head gap length
g'	Reproducing head gap length
U	Magnetic potential
φ_c	Magnetic flux in head core
R_g	Front gap reluctance
R_c	Core reluctance
i	Rear gap length
h	Core length
$A_{g'}$	Front gap cross section
A_h	Core cross section
A_i	Rear gap cross section
μ_c	Low frequency permeability of reproducing head core
μ_c'	Core permeability, real component
μ_c''	Core permeability, imaginary component
μ_f	Complex permeability at frequency f
ρ_c	Core resistivity
δ	Lamination thickness
e	Instantaneous reproducing head voltage
c	Tape coating thickness
w	Tape width
a	Head to tape separation
G	Head to tape contact length
λ	Tape recorded wavelength
v	Tape velocity
p	Tape recorded pulse width
L_c	Thickness loss in reproduction
L_a	Spacing loss in reproduction
$L_{g'}$	Gap loss in reproduction
K	Anisotropy constant

$K_0, K_1, K_2,$	Crystal anisotropy constants
K_t	Annealed induced anisotropy constant
b	Domain wall thickness
S	Particle radius
S_0	Critical particle radius for coherent rotation
d	Particle diameter
E_i	Particle internal energy
E_i'	Particle internal energy density
E_e	Particle external energy
E_e'	Particle external energy density
E_c'	Magnetocrystalline energy density
E_m'	Magnetoelastic energy density
λ_s	Isotropic saturation magnetostriction constant
D	Particle density
r_0	Minimum interparticle spacing
ρ	Apparent sample density
ρ'	Material density
T_C	Curie temperature

CHAPTER 1

INTRODUCTORY

§ 1. SCOPE OF TEXT

By way of an introduction to the subject matter of this book it is appropriate to discuss initially the scope of the contents including a description of the basic recording system to be studied. This is followed by a guide to the contents of the six subsequent chapters, from which it is hoped that readers of different backgrounds and interests may find the sections relevant to their requirements.

The rather general title of the book embraces a wider subject than is covered here. This is apparently true of most titles of technical books since definition is usually sacrificed for abbreviation. Magnetic recording can be accomplished in many ways, although two specific methods have undergone extensive development during the last decade or so. These methods may be categorized as static magnetic memories used for limited information storage in computers, where the speed of information retrieval must be high, and moving media magnetic storage devices. In the latter category the most common system in use employs a flexible magnetic tape which is transported mechanically in the vicinity of recording and reproducing transducers. This system can store unlimited quantities of information but has a relatively long access time. Although, in general, the two magnetic recording methods are quite different, they do have certain similarities. Indeed, it may be beneficial for those interested in magnetic memories, as defined above, to study the basic processes of magnetic tape recording, and vice versa, since there appears to be a growing need for devices with the combined attributes of relatively high speed access and large capacity.

Magnetic tape recording is the subject matter of this book. However, complete coverage of this subject cannot be accomplished in a monograph of this size, and this treatment concentrates on the magnetic aspects of the subject. The important mechanical problems of tape transportation and the electronic techniques for processing the signals are consequently ignored entirely: these aspects have been adequately documented in other recent engineering publications on magnetic tape recording (DAVIES [1961], STEWART [1958], SPRATT [1958] and WINCKEL [1960]). In addition to the limitation of subject matter, there is also a temporal restriction which should be appreciated. It will be quite clear, on reading published accounts of the magnetic recording process, that the analysis of the process is still under development. To a certain extent it awaits further physical insight into the basic magnetization processes occurring in permanent magnet materials. Nevertheless, advances have been made in the understanding of the fundamental processes involved and it is the purpose of this book to present a progress report on models for these magnetization processes, and on their application to the magnetic recording process. Up to the time of writing, more work has been performed on linear recording and reproduction of sine waves, due to the major initial application of tape recording to sound recording. In recent years, however, the recording of signals in the form of a coded sequence of binary pulses has considerably widened the applications of magnetic recording to storage of digital data. The present status of analysis of pulse recording and reproduction is described in this book, although it can be expected that further important work is yet to come.

§ 2. BASIC RECORDING SYSTEM

The elements of a conventional magnetic tape recording and reproducing system are shown diagrammatically in Fig. 1.1. Here a reel of magnetic tape (A) is unwound by a constant velocity tape drive (E) consisting of a rotating capstan and a pinch roller. The magnetic coating on the plastic base of the tape is thereby transported sequentially in contact with the magnetic transducers (B), (C), and (D). It is then rewound onto the take-up reel (F). The functions of the transducers are as follows: transducer (B) demagnetizes the magnetic coating, (C) records the signal by magnetizing the coating, along its length, in a manner which is some function of the time variation of the signal, while (D) detects the magnetization of the tape. Similar magnetic transduction techniques are sometimes used with magnetic storage disks and drums which are transported with constant velocity in the vicinity of the transducers, and much of the analysis of tape recording and reproducing is also applicable to such systems.

The purposes of the recording and reproducing transducers are

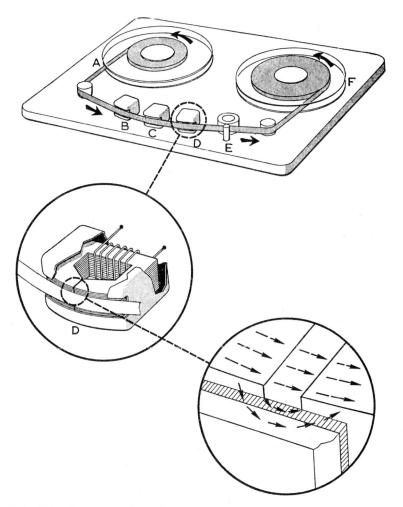

Fig. 1.1.   Representation of basic tape recording and reproducing system.

respectively to magnetize and detect as short a length of tape as possible in order that a maximum amount of information may be stored on a reel of tape. This is conventionally achieved using the leakage field at the gap in a high permeability magnetic ring core for recording, and conversely detecting the recorded magnetization pattern with a similar magnetic circuit for reproduction. A coil, wound on the ring, is used for core magnetization in recording and for core flux detection in reproduction. The design of these transducers is illustrated in the upper inset of Fig. 1.1. The lower inset illustrates the leakage magnetic flux in the gap region and the disposition of the magnetic tape. The first transducer, (B), is usually of similar design to (C) and (D) although the requirement for minimum field spread along the tape is not necessary here since its function is to erase any previous recording. Other methods of altering the state of magnetization of a permanent magnet material may be used for recording. For instance, demagnetization by a field, heat, or mechanical deformation are feasible but have not found wide application. Similarly, other methods of magnetization detection are possible but receive only brief mention in this book. The advantages of simplicity and high resolution afforded by the transducers described have largely dictated their preponderance in current magnetic recording systems.

The basic magnetic processes occurring in magnetic tape recording can be categorized as follows:

1. Magnetization of the tape by the field at the recording head gap while in contact with the high permeability poles. The magnetization acquired by an element of tape during its passage across the recording head gap will depend on the strength, direction, and duration of the field, and on the magnetization mechanisms in the material. It is desirable that the remanent magnetization of the tape be as large as possible, and for linear recording, that it be proportional to the recording field magnitude. It is not possible to produce a magnetic material having both of the above properties, and the magnetization characteristic is necessarily highly nonlinear. When amplitude linearity is required in the recording system, some

linearizing field must be added to the signal field. The function of this extra field, or bias, is to magnetize the tape in such a way that by itself it is not detectable, but that, when combined with a signal, it should produce a magnetization level proportional to that signal. The most common and successful bias is a high frequency field of sufficient amplitude to saturate the tape. The resulting tape magnetization process is similar to ideal or anhysteretic magnetization. Other biassing methods are possible, and zero bias recording is also very important for recording digital data, where linearity is unimportant.

2. Removal of high permeability poles from recorded tape. As the tape leaves the recording head, the magnetized particles no longer have their intrinsic demagnetizing fields shunted by the poles of the recording head. Consequently, these fields increase inside the particles causing reversible and irreversible demagnetization losses.

3. Natural relaxation phenomena in the magnetized tape. When the recorded tape is wound up on the take-up reel and then stored, the magnetization will re-adjust itself to the lowest energy condition for this environment. Re-adjustment will consist of self-demagnetization effects and also magnetization effects due to external fields and due to fields from adjacent layers of tape. All of these effects are normally very small in modern recording tapes.

4. Tape magnetization detection by the reproducing head gap. During reproduction, the recorded tape is transported in the vicinity of the reproducing head which detects its magnetization by a magnetometric method. The individual magnetic particles in the tape each have an associated external demagnetizing field. These fields produce the greatest magnetization of the soft magnetic core of the reproducing head when they are positioned in the region of the head gap. At this time, there is also a partial recovery of reversible demagnetization loss. If the tape is demagnetized, the particle fields are randomly directed and give rise to random magnetization of the core and a

consequent noise voltage across the core winding. On the other hand, if the tape is recorded, the particle fields become ordered in direction and produce a head core magnetization proportional to the number of unidirectionally magnetized particles.

The contents of the following six chapters are devoted to a detailed analysis of the magnetic processes described above, including a study of the properties of heads and tapes. A magnetic recording and reproducing system employing these processes and components is described in Chapter 7.

The recording process is described in Chapters 2 and 3, and this includes all changes in the tape magnetization which contribute to the magnetization pattern of a recorded tape at the point where the tape leaves the vicinity of the recording head: thus, items 1 and 2 above are considered. Chapter 2 deals with the ac bias recording process. Those readers who are only interested in the physics of this process may refer to §§ 4–6, where an analysis of the anhysteretic magnetization process is modified to account for the boundary conditions occurring in recording. It should be appreciated that at present the anhysteretic magnetization process is not completely understood in tape materials, since knowledge of the precise magnetization mechanisms in tape materials is incomplete. The magnetization processes occurring in permanent magnet powders are described in Chapter 5 and the properties of practical materials in Chapter 6. Models have been used to describe the magnetization processes and their limitations should be borne in mind. For those readers initially requiring a simplified picture of the various recording processes, a model is developed in Chapter 2, § 2, for ac bias and in Chapter 3, §§ 1.1 and 2.1 for zero and dc bias respectively. Used cautiously, this model can clearly describe many aspects of the recording process, since it takes account of the finite dimensions of the tape, the finite extent of the recording zone and the non-linear magnetization characteristic of the medium. It does not, however, adequately describe the magnetization mechanism in tapes and cannot account quantitatively for recording properties. Chapter 3 deals with sine wave and pulse recording without bias and also with the lesser used method of dc bias. Finally, other possible techniques for recording are described which use heat as the linearization agent.

Chapter 4 covers item 4 above in a comprehensive description of the reproducing process. Reproduction of all tape magnetizations are dealt with and this includes recorded signals, noise, and printed signals. The print-through effect itself belongs, of course, neither to the recording nor the reproducing process. The basic phenomena of magnetic relaxation effects, which give rise to print-through, are considered in Chapter 5, § 2.4, whereas the practical printed magnetizations and consequent reproduced voltages are included in Chapter 4, § 7.

Looking at the contents from the point of view of basic magnetization processes, the magnetization mechanisms in small particles are described in Chapter 5. The particles of greatest interest in tapes are single domains, although any practical powder will additionally contain particles too small or too large to be single domains; these are paramagnetic and multidomained respectively and are also described in Chapter 5. The time stability of magnetization is another basic process included in this chapter. When single domain particles are packed together in a tape, their external fields interact with neighbouring particles, tending to reorient the particle magnetization directions. This interaction plays an important role in the anhysteretic magnetization process described in Chapter 2, §§ 4–6. Demagnetization processes are active in other phases of the recording and reproducing process, and a magnetized tape misses no opportunity to change towards the disordered state. Firstly, the self-demagnetizing field of a single domain particle usually depends on its direction of magnetization: this is considered in Chapter 5. Demagnetization effects also occur due to the finite recorded wavelength. This effect is largest for short wavelengths and causes both reversible and irreversible losses when the recorded tape leaves the vicinity of the head, as discussed in Chapter 2, § 6. Some of the reversible demagnetization loss is recovered on reproduction and this topic is discussed in Chapter 4, § 2.4.

§ 3. UNITS AND SYMBOLS

During the course of the literature study for this book, it became apparent that the question of preferred units is still unresolved. However, in this work emphasis is placed on the physics of the magnetization processes involved in magnetic recording and this has determined the use of the C.G.S. electromagnetic system of units (see BROWN [1960b]).

The symbols employed in this book to denote the magnetic characteristics of materials follow those in common use in related current literature. In order to emphasize the link between recording processes and basic magnetic characteristics, a similar symbol structure is used to describe recording fields and recorded magnetization. Intensity of magnetization is denoted by the symbol I, and the apparent magnetization of powders and films, used for magnetic recording, by I''. For the important symbols used in the text see page xii.

CHAPTER 2

MAGNETIC RECORDING PROCESS – AC BIAS

§ 1. INTRODUCTION

The magnetic recording process employing a high frequency magnetic field together with the signal field is the most common and successful technique for analogue recording in use today. Although this technique is relatively easy to operate in practice, it is difficult to account quantitatively for the magnetization processes involved. The problem of describing adequately the magnetization process in magnetic tapes is still being actively analyzed, and it is only recently that the underlying physical principles have been reasonably well understood. Consequently, up to the present time, the ac bias recording process has been described in terms of models. The usefulness of such models is that they can clarify certain aspects of the recording process for a particular set of recording conditions. It is important, however, to recognize the specific limitations, due to assumed simplifications, in order to avoid erroneous conclusions. Apparent contradictions between models which are inadequate, rather than incorrect, can occur if the limitations are not clearly defined.

It is natural that the well-known remanent magnetization characteristic of the recording material was used in early models for the recording process, e.g. HOLMES and CLARK [1945], CAMRAS [1949], ZENNER [1951]. However, under practical recording conditions this characteristic is inadequate, since the ac bias reverses the recording field polarity several times during the time that a tape element is in the recording field zone. To describe the resulting magnetization changes adequately it is then necessary to consider the combined effect of a number of asymmetrical magnetization cycles. A straightforward but laborious approach would be to compute the hysteresis cycles experienced by

9

each element of tape due to the combined bias and signal fields. Fortunately, however, the combined effects of ac and dc fields give a magnetization which is characteristic of the material for conditions which are rather similar to those occurring in recording. This process, known as anhysteretic magnetization, comprises the simultaneous application of a constant dc field and an ac field which is gradually reduced from a saturating value to zero. As will be described later, the anhysteretic remanent magnetization is essentially proportional to the applied dc field for magnetization levels up to about half the maximum value. This linear function, extending to a greater or lesser relative magnetization level in all permanent magnet materials, is the basis of the linearizing process occurring in ac bias recording. The magnetization process experienced by a tape element during recording differs from the ideal anhysteretic magnetization process in several important respects: for instance, the dc field is neither constant nor unidirectional during the reduction of the ac field as the element leaves the recording zone. However, a sufficient number of ac bias cycles is experienced while the tape element is leaving the recording zone for anhysteresis to occur, and modifications for practical conditions can be accounted for. The major aim of the following development of the ac bias recording process is, first, to account for the anhysteretic magnetization process in terms of the internal magnetization processes in the tape coating. This in itself is a complicated process and formal models are used prior to a simplified quantitative analysis. Secondly, it is necessary to take account of the modifications due to variations in the applied field intensity, relative phase, and direction for all the tape elements, in order to compute the final tape magnetization.

The development of the recording process as outlined above is preceded by an initial description in terms of the well-known magnetization characteristics. A set of simplifying conditions is used which allows some of the commonly observed recording characteristics to be portrayed qualitatively in terms of the recording field experienced by the complete coating. In particular, those effects are emphasized which are due to the non-uniform recording field amplitude at different depths in the coating. Although a non-linear magnetization characteristic is assumed in this simple model, it is too crude to account quantitatively for the recording characteristics. Consideration of experimental magnetization loops then leads naturally to the study of anhysteresis.

§ 2. A SIMPLE AC BIAS RECORDING MODEL

Since the ac bias recording process is complicated, a rather lengthy description is necessary to take into account all of the active factors in the process. Before embarking on a study of the detailed magnetization mechanisms in the ac bias recording process, described in subsequent sections, the reader who is not familiar with magnetic recording may benefit from the following simplified model which describes qualitatively the essentials of the process (BAUER and MEE [1961]). The model chosen takes into account the non-linear magnetization characteristics of permanent magnet materials and also the important contours of the recording field configuration for long wavelength recording. Since the basic recording mechanism is described satisfactorily by this model, it will be used as an introduction for each of the biassing techniques described.

It is assumed that the tape coating consists of a uniform distribution of identical magnetic particles, each being magnetically isolated from its neighbours. Moreover, it is assumed that the tape responds only to the longitudinal component of the applied field and that its magnetization loop is the same as those of the individual particles; i.e. rectangular. These conditions are fulfilled, in part, by present day tapes consisting of needle shaped single domain particles oriented in the longitudinal direction of the tape. Ideally, such particles have rectangular hysteresis loops when magnetized along their major axes. Practical tape coatings will depart from ideal, since variations in particle size and shape, as well as clumping, result in a spread of critical fields for switching the magnetization of the particles. In addition, the probability exists that the fields of neighbouring particles cause sufficient particle interaction to modify the local fields acting when an external signal field is applied. The effects of interactions and particle variations are significant in tape recording, as will be described later.

The recording field considered here is the component of the external field existing parallel to the direction of tape motion over an infinitesimal gap between pole pieces of infinite permeability. Fig. 2.1 shows the tape motion and gap direction referred to x, y coordinates existing parallel to the tape motion and perpendicular to the head-tape contact plane, respectively. The field (H) produced at a distance (R) from the gap by a current (i) in the winding around head core is given by

$$H = 4ni/R . \tag{2.1}$$

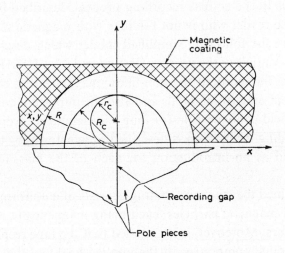

Fig. 2.1.   Magnetic recording model.

The component of this field parallel to the tape motion (x axis) is

$$H_x = 4niy/(x^2 + y^2) . \tag{2.2}$$

Assuming a single value of the critical field (H_c) for magnetization of the particles, the locus of the boundary for which $H_x = H_c$ is given by

$$\left.\begin{array}{l} x^2 + y^2 = 4niy/H_c , \\[2mm] x^2 + (y - r_c)^2 = r_c{}^2 , \text{ where } r_c = 2ni/H_c . \end{array}\right\} \tag{2.3}$$

or

This is a circle of radius r_c with centre at the point $(0, r_c)$ (see Fig. 2.1). The applied field inside the circle, being greater than H_c, is sufficiently

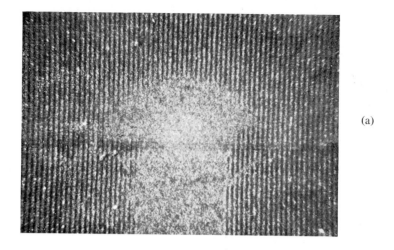

(a)

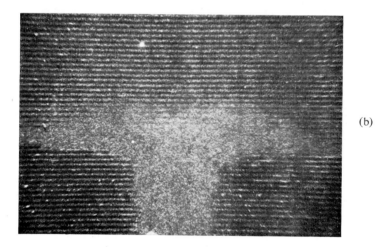

(b)

Fig. 2.2.   Gap edge erasure contour; acicular particle tape.
a. Gap length parallel to longitudinal tape direction.
b. Gap length perpendicular to longitudinal tape direction.

large to magnetize completely the magnetic material, whereas outside the circle it is less than H_c and the magnetic material remains unmagnetized. The diameter of the magnetization circle $R_c(x = 0)$ is given by

$$2r_c = R_c(x = 0) = 4ni/H_c = 4n(i_b + i_s)/H_c, \qquad (2.4)$$

where i_b and i_s are bias and signal currents, respectively.

Eq. (2.4) indicates that the diameter of the magnetization circle is proportional to the sum of the signal and bias currents. Thus, the magnetization circle can be visualized as a recording cylindrical volume whose size varies in accordance with the applied signal for constant bias. Using a pre-recorded tape, the effective field contour around a finite gap in a recording head may be studied by placing the tape in contact with the gap which carries sufficient ac flux to erase the recording in the gap region (GUCKENBURG and MEE [1961]). The erased region may then be studied by making the residual recording visible using the magnetic ink technique. Figs. 2.2a and b show the erased zones for the particle orientation direction parallel and perpendicular to the conventional direction of tape motion, respectively. It can be seen in Fig. 2.2a that an approximately circular erased zone is obtained for the conventional recording situation in conformity with the assumptions of the model. For an infinitely small gap, the perpendicular equal field boundary would be two semi-circles whose diameters lie along the gap edge and which touch each other at the gap. This condition is approached in Fig. 2.2b.

The model of the recording process thus consists of a cylindrical volume of longitudinal magnetization, tangent to the recording head gap, penetrating the moving tape with an instantaneous diameter proportional to the total instantaneous recording current. The model is primarily applicable to long wavelength recording phenomena where the magnetization acquired by the surface layers is not overriding. However, for tape layers at a distance from the head which are small compared to the gap length, the effect of the perpendicular field cannot be ignored.

Eq. (2.4) indicates that the recorded magnetization is here proportional to the recording head current. Of course, this is an extreme case and the degree of linearization achieved with finite gap lengths will depend on the decrement of the recording field through the coating, that is, on the ratio of the coating thickness to the gap length (see Ch. 3 § 1.1). However, in modern high resolution recording systems, the coating thickness c may be several times greater than the recording

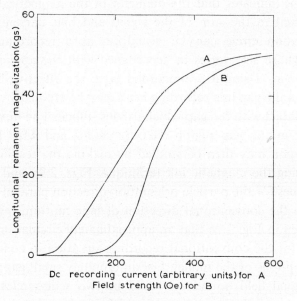

Fig. 2.3.   Remanent magnetization *vs.* dc field.
A. Narrow gap recording head field, B. Solenoid field.

head gap length g, and considerable linearization then occurs. This is illustrated in Fig. 2.3 where the remanent magnetization characteristic (curve B) is compared with the recording characteristic for the same tape when the ratio $c/g = 3$.

Using the recording model, the ac bias recording process consists of laying down successive overlapping cylindrical magnetization volumes of opposite polarity. In the absence of a signal field, the ac bias field creates equal maximum cylindrical volumes leaving equal volume magnetization shells in the tape after it has passed the recording head.

This condition is shown on the left hand side of Fig. 2.4. On adding a signal field, low in frequency with respect to the bias field, the overlapping of adjacent cylinders is no longer symmetrical and the resultant area of tape magnetized in the signal field direction increases. Provided the remote side of the tape is not reached during the recording

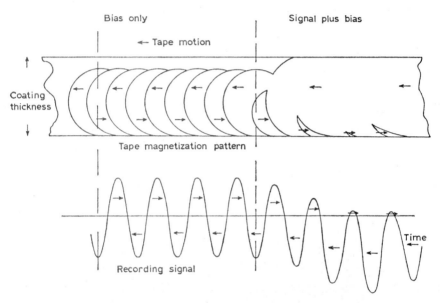

Fig. 2.4.   Ac recording signal and tape magnetization pattern.

process, the increase of magnetization area in the signal field direction is proportional to the field magnitude. Penetration of the remote side of the magnetic coating by the recording field leads to distortion.

In an analogous fashion, the tape magnetization for a given signal depends on the magnitude of the ac bias and increases with bias amplitude until the cylindrical magnetization zone reaches the remote side of the tape. Increasing the bias amplitude will push this area beyond the coating, resulting in loss of signal field magnetization. In this way the maximum in the recording sensitivity curve, shown in Fig. 2.5, as a function of bias amplitude, is explained.

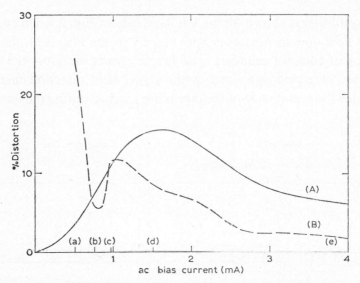

Fig. 2.5.   Output and distortion *vs.* bias for long wavelength recording.
(A) Reproducing head output (arbitrary units).
(B) Total harmonic distortion.

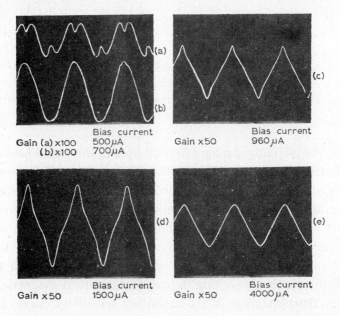

Fig. 2.6.   Distortions occurring in ac bias recording.

The model may be extended to explain the dependence of harmonic distortion content on the strength of the ac bias, in terms of the boundary conditions set by the front and back surfaces of the tape, as the magnetization cylinder grows into and beyond the tape coating

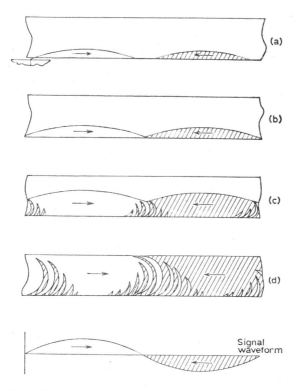

Fig. 2.7.   Long wavelength recording for different bias levels.

(MEE [1962]). The total harmonic distortion is plotted in Fig. 2.5 as a function of ac bias level using a narrow gap recording head and a long recorded wavelength compared to the coating thickness. The distorted waveforms obtained at the reproducing head for various bias levels are shown in Fig. 2.6a–e and the corresponding bias currents are noted on Fig. 2.5. The distortions are those due to a signal level large enough to approach tape saturation when the bias level is adjusted for maximum output. For smaller signal levels, a similar set of reproducing head waveforms is obtained, except that the saturation distortion at the higher bias levels is absent.

All of the distortions observed may be qualitatively explained by considering the magnetization patterns in terms of the model. The patterns corresponding to the distortion waveforms of Fig. 2.6 are shown in Fig. 2.7. The magnetization pattern with 500 μA bias (Fig. 2.7a) is a variable depth magnetization, following the signal amplitude, due to the growth of the cylindrical magnetization volume into the coating. This is similar to a zero bias recording since the bias component of the magnetization cylinder is not sufficient to penetrate the tape. However, on increasing the bias amplitude (Figs. 2.6b and 2.7b) a point is reached where the bias is just sufficient to magnetize the surface layer. At this point the signal recording is pushed into the tape far enough to avoid the distortion at the low signal part of the waveform and a distortionless recording results. Distortion rapidly rises again on further increase of bias amplitude (Figs. 2.6c and 2.7c). It can be seen from Fig. 2.7c that this is due to the sinusoidal magnetization being carried further into the tape coating by the bias, giving larger maximum flux even though the tape magnetization returns to zero each time the signal passes through zero. The rate of change of recorded flux is thus greater at zero signal than that corresponding to a sinusoidal curve, leading to sharply peaked reproducing head output waveform (Fig. 2.6c). When the magnetization due to the bias extends through the whole coating thickness, depth recording by the superimposed signal ceases and the overall tape magnetization change is due to varying degrees of overlap of the bias cylinders as shown in Fig. 2.7d (and on the right hand side of Fig. 2.4). In Fig. 2.7d the maximum of the signal is large enough to produce complete tape saturation, thus giving the distorted waveform of Fig. 2.6d. Finally, if the bias is substantially increased, this distortion disappears due to the relatively smaller change of magnetization cylinder diameter with signal amplitude, resulting in non-saturation of the tape.

Summarizing, the simple model of the recording process takes into account the important property of the recording field configuration for long wavelength recording, that is, the 'y' decrement of the field magnitude through the coating thickness; for long wavelengths, the 'x' decrement in the direction of tape travel is not significant. As a consequence, those recording characteristics depending on boundary effects, such as the finite coating thickness, are also adequately explained. In this way, the recording sensitivity and distortion as a function of bias amplitude have been described in terms of the growth

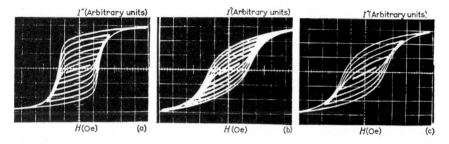

Fig. 2.8.   Family of magnetization *vs.* applied field characteristics for oriented acicular particles of γ Fe$_2$ O$_3$. ($H_{max} = 1000$ Oe.)
a. Along orientation direction.
b. Perpendicular to (a).
c. Unoriented acicular particles.

of the cylindrical magnetization zone into, through, and beyond the tape coating. However, quantitative application of the model is not possible due to over-simplified expression of the recording field configuration and the tape magnetization characteristics. The ideal rectangular hysteresis loop assumed in the model is that proposed some years ago by STONER and WOHLFARTH [1948], where it is assumed that each particle is identical and magnetically isolated from its neighbours. In this event, as has been shown, the increase of tape magnetization with applied signal field is due to an area modulation of the magnetization zones created by the ac field. Whereas this mechanism undoubtedly exists, it is not the controlling factor in determining the recording sensitivity. As will be described in the following development of the ac bias recording process, internal fields acting on the magnetic particles of the tape modify their effective coercivities, or switching fields, and play a cardinal role in determining the anhysteretic magnetization characteristic. In turn, these internal fields therefore control the response of the tape material to the recording field.

§ 3. DETAILED STUDIES OF HYSTERESIS LOOPS

It has been pointed out that the simple model of non-interacting particles is inadequate when quantitative evaluation of the anhysteretic

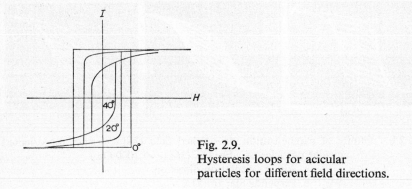

Fig. 2.9.
Hysteresis loops for acicular
particles for different field directions.

magnetization characteristic and of the ac bias recording characteristic are required. The practical hysteresis loops for tape powders will therefore be studied in more detail in an attempt to obtain a satisfactory analysis of the operative magnetization processes.

Figs. 2.8*a* and *b* show a family of hysteresis loops for a tape consisting of oriented acicular particles of ferric oxide measured in the orientation direction and perpendicular to it, respectively. By comparison, Fig. 2.8*c* shows the loops corresponding to an unoriented sample. The rectangular hysteresis loop, assumed for the previous simple recording model, applies only to those particles in the tape which are magnetically isolated from their neighbours, and which are perfectly aligned with the applied field direction. If the particles are not aligned parallel to the recording field, their magnetization loops are, in general, not rectangular and their remanent magnetizations and coercive forces are reduced. The detailed magnetization mechanisms

involved are fully described in Chapter 5. For the present, the magnetization will be assumed to rotate uniformly inside a particle (STONER
and WOHLFARTH [1948]), and the hysteresis loops as a function of angle
between the particle and the applied field are then as shown in Fig.
2.9. In addition to the effects of angular dispersion of the particles in a
practical magnetic tape, a distribution of particle shapes and sizes will
also lead to a spread of the particle switching fields. Nevertheless, in
modern magnetic tapes, good particle size and shape uniformity is
achieved, along with efficient orientation of the particles, leading to
hysteresis loops having the high degree of rectangularity shown in
Fig. 2.8a. Thus, no major modification of the recording model is
necessary to take into account the effect of practical particle switching
field distributions. The magnetization will fall off less rapidly at the
boundaries of the recording zone, but no significant change of recording sensitivity would result. In effect, the resolution of the recording of
the ac bias waveform is impaired by considering a distribution of
particle switching fields, and the distinct recording of the bias field
pictured in Fig. 2.4 would not occur. This, however, does not detract
from the basic validity of the model.

From Fig. 2.8a and c it is seen that the major changes in the hysteresis cycle due to particle alignment are an increase of remanent
magnetization and a smaller dependence of coercivity on the maximum
field strengths of the minor loops. The Stoner–Wohlfarth magnetization model (see Ch. 5), which describes the hysteresis loop as a function of particle demagnetization energy changes during the rotation of
the direction of magnetization, is able to explain broadly the loop
changes described; however, the relatively constant magnitude of
coercivity with particle orientation shown by Figs. 2.8a and b is not
satisfactorily explained, and other magnetization mechanisms must be
invoked to do so (see Ch. 5).

Unfortunately, it is not sufficient to take account of particle switching field distribution alone in order to describe fully the magnetization
characteristics of a tape powder. The saturation hysteresis loop for
oriented acicular particles of γFe_2O_3 is shown in Fig. 2.10a along with a
family of minor loops, about positive saturation, for different values

of negative maximum field $(-H_{\max})$. Similar loops for spherical particles controlled by crystal anisotropy are shown in Fig. 2.10b. As in the case of symmetrical minor loops, the portions of the loops from $-H_{\max}$ to zero field, representing reversible magnetization changes, have equal differential susceptibilities (χ_d'') for all values of $-H_{\max}$. However, for positive fields of about $H = 0.2H_{\max}$, the differential susceptibility suddenly increases for all the minor loops due to the onset of irreversible changes and its magnitude increases with the size

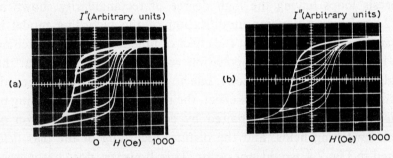

Fig. 2.10.   Hysteresis loops for tape powders:
a. Oriented γ Fe$_2$ O$_3$
b. Cobalt iron oxide
$H_{\max} = 1000$ Oe.

of the minor loop. The consequence is that the minor loops are asymmetrical with respect to the magnetization axis as well as the field axis. Now, in the anhysteretic magnetization process, and the ac bias recording process, minor loops are traced by the slowly diminishing ac field and these will be asymmetrical with respect to the field axis if a dc field or signal is superimposed. Based on these observations a model of the minor loops traced during anhysteretic magnetization is shown in Fig. 2.11, ignoring reversible magnetization changes. In the model, the irreversible differential susceptibility is made to be proportional to the magnetization. A most important result is that the remanent anhysteretic magnetization $[I_{ar}''(H_{dc})]$ is less than the maximum value. On the other hand, for the simple non-interacting particle model initially assumed, χ_d'' would be independent of magnetization level and the anhysteretic remanent magnetization would have the maximum value for all finite dc fields. It is now known that interaction fields between particles are the cause of the asymmetrical minor loops, and a description of the anhysteretic magnetization process, which takes the interaction fields into account, is given in the next section.

With regard to the simple recording model, the number of particles

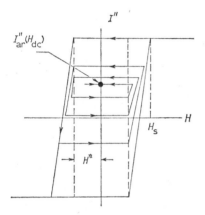

Fig. 2.11.
Hysteresis loops traced during anhysteretic magnetization process.

in a demagnetized tape which can be switched by an applied field is the same, with or without accounting for interaction. However, the critical fields for switching those particles initially magnetized oppositely to the applied field are modified by the interaction fields and a range of effective switching fields results. Thus the magnetization is not uniform inside the recording zone, and the cylindrical magnetization zone concept becomes invalid if the magnetization is assessed from the volume of the cylindrical shell patterns after recording. Since the effective interaction fields are like a dc bias field opposing the applied field, the recording sensitivity would be lower than if interaction was negligible. A detailed description of the ac bias recording process must then take into account interaction fields. In the following sections the anhysteretic magnetization process is first described in detail, followed by modifications which occur in ac bias recording.

§ 4. ANHYSTERETIC MAGNETIZATION PROCESSES

The conclusion of the foregoing study of the ac bias recording mechanism is that the basic process involved is a modification of the ideal anhysteretic magnetization process. Furthermore, when recording with a ring head, significant variation of the applied field amplitude occurs for different layers in the tape coating causing additional modification of the ideal anhysteretic magnetization characteristic. The simple recording model described in § 2 takes into account a field gradient through the coating and describes the modified anhysteretic magnetization process for an idealized magnetization model, ignoring interaction effects between particles. However, the latter are of prime importance in determining the anhysteretic magnetization characteristics. In the following three sections of this chapter the ac bias recording process is to be developed in more detail along the following lines:

§ 4. Description of anhysteretic magnetization processes including magnetization dependent internal field effects.

§ 5. Anhysteretic magnetization processes including magnetization independent internal field effects.

§ 6. Application of anhysteretic magnetization processes to ac bias recording.

§ 4.1. *Ideal Anhysteretic Magnetization*

Magnetization achieved by the ideal anhysteretic method is determined by the magnitude of the applied dc field (H_{dc}) irrespective of any previous magnetic history. To obtain this condition, a large symmetrical alternating field is superimposed on the applied field. The amplitude of the alternating field (H_{ac}), which is initially sufficient practically to saturate the material, is gradually reduced to zero. Early work on this method of magnetization showed that the anhysteretic magnetization curve had no point of inflection corresponding to the normal magnetization curve and that it is always concave with respect to the field axis. The anhysteretic remanent magnetization $[I''_{ar}(H_{dc})]$ is obtained by subsequent removal of the applied dc field. In magnetic tapes, due to the small reversible magnetization component, the remanent magnetization curves are similar to the magnetization curves; in this section

remanent magnetization curves will be studied. The initial susceptibility obtained by anhysteretic magnetization can be many times greater
than that obtained without the alternating field. Fig. 2.12 shows
anhysteretic remanent magnetization curves along the direction of
orientation of an acicular iron oxide particle tape as a function of the
ac and dc fields applied.

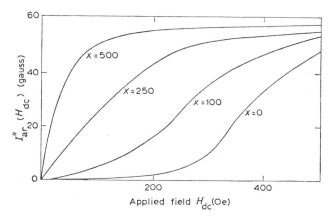

Fig. 2.12.   Anhysteretic magnetization curves.
$x = $ peak H_{ac} (Oe).

In Fig. 2.12 it is seen that the remanent magnetization curve for zero
ac field rises rapidly between applied dc fields from 250 to 450 Oe. It
is assumed that in this range the applied field is sufficiently large to
overcome the effective switching fields of the particles and that the
steepness of the curve is a function of the number of particles switching
their magnetization towards the applied field direction. When an alternating field is superimposed on the dc field, equivalent magnetization
switching occurs when the same total field is applied. Thus, for large
ac fields, small dc fields are required to produce irreversible magnetization on reducing the ac field to zero.

When the initial ac field is large enough to switch the magnetization
of all the particles, a limiting condition is reached (Fig. 2.12, $H_{ac} > 500$
Oe). As the ac field is reduced, the critical switching fields for the

hardest particles will be reached first, and at this point will cease to follow the alternating field and become frozen in the applied field direction. Thus, as the ac field is reduced further, a unidirectional magnetization grows at the expense of the alternating magnetization at a rate depending on the number of particles whose switching fields lie in the range of instantaneous applied field. However, it is a condi-

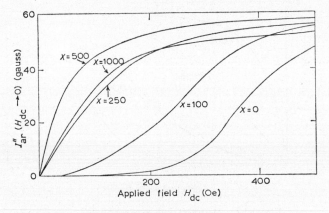

Fig. 2.13.   Modified anhysteretic magnetization curves.
H_{ac}/H_{dc} = constant.
x = peak H_{ac} (Oe).

tion of true anhysteretic magnetization that the ac field decrement is slow enough to ensure that the negative peak of the total signal $(-H_{ac} + H_{dc})$ is followed by a positive peak of equal or greater amplitude. Thus, if the anhysteretic magnetization process were completely described by the above procedure, the anhysteretic remanent magnetization curve would rise with infinite slope to the maximum value. The finite values obtained in practice are due to internal fields which must be taken into account.

§ 4.2.   Modified Anhysteretic Magnetization

An important difference between the ideal anhysteretic magnetization process and that occurring in ac bias recording is that, in the

latter case, the dc field (or signal field) is not constant during the magnetization process but diminishes at the same rate as the high frequency
field. The effect of this modification is to reduce the magnetization for
a given initial dc field by an amount which depends on the initial value
of the ac field. If the ac field is very large, then the dc field will have
reduced considerably at the time that the ac field has fallen to the

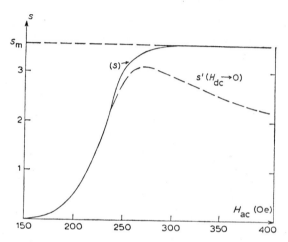

Fig. 2.14.   Initial anhysteretic susceptibility, s, $vs.$ maximum ac field (H_{ac}).

switching field range of the material. On the other hand, for small ac
fields, that is if $H_{ac\ max}$ is not large enough to reorientate the magnetization vectors of all the particles, the difference between the anhysteretic remanent magnetization acquired by the two methods (H_{dc}
constant or H_{ac}/H_{dc} constant) will reduce. Experimental anhysteretic
remanent magnetization curves for iron oxide powders with ac and dc
fields falling together are shown in Fig. 2.13 and are in general agreement with the foregoing description of these magnetization processes.
   The initial anhysteretic susceptibility, as a function of the ac field
amplitude, is shown for the two methods in Fig. 2.14. Increasing the ac
field leads to increase of initial anhysteretic susceptibility, s, until a
maximum is reached (Fig. 2.14). In the case of ac and dc fields falling
together, however, the initial susceptibility decreases at high ac fields

since the ac field is then large enough to magnetize all of the particles and the remanent magnetization acquired $I''_{ar}(H_{dc} \rightarrow 0)$ depends only on the ratio of dc to ac field. As will be seen later, the peak in the initial susceptibility curve gives rise to an optimum value of bias field for maximum recording sensitivity.

§ 4.3.   *Effect of Internal Fields*

The ideal and modified anhysteretic magnetization mechanisms thus far outlined lead to infinite initial anhysteretic susceptibility, whereas the experimental curves of Fig. 2.14 indicate that a maximum value of only about 3 is obtained in γFe_2O_3 powder. Internal fields are responsible for the finite values obtained, and these will be considered in two formal ways. Firstly, it will be assumed that an internal field exists which is proportional to the mean value of the magnetization. In the next section, anhysteresis is considered in terms of a distribution of internal fields due to interaction between the particle demagnetizing fields and the neighbouring particles; it is then assumed that a distribution of interaction fields exists which is independent of the state of magnetization of the material.

The effect of an internal field which is proportional to the average magnetization level on the anhysteretic magnetization process will be to modify the process and produce an apparent initial susceptibility s' where

$$s' = [I'_{ar}(H_{dc})]/H_{dc}, \qquad (2.5)$$

where $I'_{ar}(H_{dc})$ is the anhysteretic remanent magnetization acquired in the presence of an internal field. If the internal field is a demagnetizing field, it will oppose the static and alternating components of the applied field. The latter effect is considered to be negligible since the initial amplitude of the alternating field is sufficient to saturate the material. The static field is then opposed by a demagnetizing field H_d given by

$$H_d = N_e[I''_{ar}(H_{dc})], \qquad (2.6)$$

where N_e is the external demagnetizing factor and $I''_{ar}(H_{dc})$ is the instantaneous value of the static magnetization acquired. Thus the acquisition of remanent magnetization during the run-down of the ac field is controlled by the instantaneous value of the demagnetizing field. The true anhysteretic susceptibility is then obtained by integration of the increments of magnetization divided by the instantaneous field, H_{dc}-H_d (NÉEL [1943]).

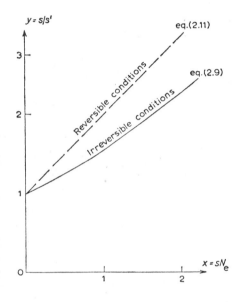

Fig. 2.15. Theoretical dependence of apparent initial anhysteretic susceptibility on the external demagnetization factor.

Hence

$$s = \int_0^{I'_{ar}(H_{dc})} \frac{\delta I''_{ar}(H_{dc})}{H_{dc} - N_e I''_{ar}(H_{dc})}, \tag{2.7}$$

leading to

$$\exp(-N_e s) = 1 - \left[\frac{N_e I'_{ar}(H_{dc})}{H_{dc}}\right]. \tag{2.8}$$

Combining with eq. (2.5) gives

$$s' = [1 - \exp(-N_e s)]/N_e , \tag{2.9}$$

where s' is the anhysteretic susceptibility in the presence of a demagnetizing field.

Eq. (2.9) is plotted in Fig. 2.15 alongside the corresponding relation assuming the magnetization to have been acquired reversibly. In this case,

$$I'_{ar}(H_{dc}) = s \left[H_{dc} - N_e I'_{ar}(H_{dc}) \right] \tag{2.10}$$

leading to

$$s' = I'_{ar}(H_{dc})/H_{dc} = s/(1 + sN_e) . \tag{2.11}$$

Experimental measurements of s' on permanent magnetic alloys with an external demagnetizing field acting (NÉEL [1943]), give good agreement with eq. (2.9) and thus support the analysis of anhysteretic magnetization based on irreversible magnetization effects. Thus, the effect of a large demagnetizing field is to reduce the initial anhysteretic susceptibility to a finite value which is the reciprocal of the demagnetizing coefficient (eq. (2.9)). If the switching fields of the particles are uniformly distributed over a small range, the effect of the demagnetizing field on the anhysteretic remanent magnetization curve is to shear it to a slope $1/N_e$ so that it rises linearly to the maximum value. If, on the other hand, an internal self-demagnetizing coefficient N_i is overriding and the external coefficient negligible, then $s' \rightarrow 1/N_i$. Of course, the internal fields could be represented by a magnetizing rather than demagnetizing field. However, in the absence of other demagnetizing fields, this would result in infinite initial anhysteretic susceptibility which is contrary to experimental results.

The result of this study of magnetization dependent internal fields is that finite initial anhysteretic susceptibility can be accounted for by assuming an internal demagnetizing field. For a single amplitude demagnetizing field the anhysteretic magnetization curve would rise with constant slope to the maximum value. Practical anhysteretic magnetization curves, however, depart from linearity at relatively low magnetization levels as illustrated in Fig. 2.20. In principle it is possible to account for this non-linearity in terms of a spread of internal demagnetization factors (WOHLFARTH [1960]). However, the major problem in assigning the observed anhysteretic magnetization characteristics to magnetization-dependent internal fields lies in determining the physical origin of the fields. Interparticle interaction fields appear to be the logical physical basis for internal fields. The net direction of the internal field with respect to the applied field direction depends critically on the geometrical arrangement of the particles, as has been shown by considering certain restricted arrays of particles (WOHLFARTH [1955]). The possibility of either an effective demagnetizing or an effective magnetizing internal field, depending on the symmetry of the particle arrays, may account for some of the observed variations of initial anhysteretic susceptibility between different permanent magnet materials. However, in general, for a random distribution of single domain particles, it is to be expected that the local internal field is the summation of randomly directed fields from neighbouring particles which remain randomly directed with changes of macroscopic magnetization. In this event the internal field is magnetization-independent. The effect of particle interaction fields on the anhysteretic magnetization characteristic is further analyzed in the following section.

§ 5. PARTICLE INTERACTION EFFECTS

The significance of interparticle interaction fields in tape powders may be demonstrated by observing the change of bulk coercive force on dilution of a powder sample (see Ch. 6, § 1.1). Even more significantly, the coercive force of a single isolated particle of γ Fe_2O_3 has been measured (MORRISH and YU [1956]) and a switching field of 800 Oe was observed, which is about three times the coercive force of similar particles in magnetic tape. Direct evidence of the existence of an effective inter-particle field has been obtained in oriented acicular iron particles (CRAIK and ISAAC [1960]). Here it has been found, by use of powder patterns, that regions in the material exist where all the particles are magnetized in the same direction. These regions have been termed 'interaction domains'.

The detailed process of the influence of particle interaction fields on the magnetization loops of neighbouring particles is complicated and has not, as yet, received full analytical expression. The magnitude and direction of the interaction field at any point will depend on the local packing density and the directions of magnetization of the particles. Consequently, if the magnetization of one particle rotates under the

influence of an external field, the local field in its vicinity will change
and the effective magnetization loops of neighbouring particles will be
affected. In this way, the composite magnetization loop for a pair of
interacting particles will depend on the relative switching fields of the

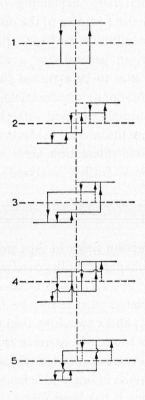

Fig. 2.16. Hysteresis loops of
interacting rectangular loop
particle pairs (NÉEL [1958]).

particles, assuming rectangular intrinsic hysteresis loops, and the sign
and magnitude of the coupling field. Various possible forms of the
composite hysteresis loops for interacting pairs are shown in Fig. 2.16
(NÉEL [1958, 1959]). Forms 3 and 4 correspond to particles with weak
interaction fields ($\approx H_c$), as may be found in tapes. The remanent
magnetization intensity is positive (Form 3) or negative (Form 4),
depending on the relative magnitudes of the saturation intensities, I_{s1}
and I_{s2}, and coercivities H_{c1} and H_{c2} of the particles. In addition, the

magnitude of the remanent magnetization I_r is given by the addition or subtraction of I_{s1} and I_{s2}. In this way I_r can be zero if $I_{s1} = I_{s2}$ for some particles in weak interaction. This model, therefore, introduces the additional variable of a distribution of remanent magnetization values for interacting pairs.

The hysteresis loops formed by the interacting pairs model may be regarded as a combination of symmetrical and asymmetrical loops. From Fig. 2.16 it can be seen that, for small fields applied to a sample consisting of all six forms in the demagnetized state, magnetization will take place initially in those forms for which the small asymmetrical loops are distributed about $H = 0$. Further sophistication of the interacting pairs model is required to account for practical interacting conditions involving many particles.

§ 5.1.   *Dipole Field Model*

In order to facilitate the calculation of the interaction fields for a system of interacting particles, it may be assumed that the dispersion of particles is sufficiently dilute and uniform for the fields to be those of magnetic dipoles. Another possible approach, if close proximity is obtained, is to consider only the field of the nearest neighbouring particle. The former approximation leads to an expression for the rms interaction field $\tilde{H}_i$ due to a random array of dipoles of average moment, $\bar{m}$ (ELDRIDGE [1961])

$$\tilde{H}_i = 2.9 \, I_s/D^{\frac{1}{2}}r_0^{\frac{3}{2}} . \tag{2.12}$$

where I_s is the saturation intensity of magnetization for the particle array, D is the particle density and r_0 the minimum interparticle spacing.

It is assumed that the interaction field on a particle shifts the intrinsic rectangular hysteresis loop along the H axis by an amount equal to the component of the field parallel to the particle axis. The perpendicular component may be ignored since it produces no asymmetry of irreversible magnetization. Changes of particle coercivity also occur under such interaction conditions (BROWN [1962]); these will produce some modification of the simple model described but are ignored in this treatment. When the particle assembly is magnetized anhysteretically, the magnitude of the component of the dc field, H_{dc}, along the particle axis with respect to the parallel component of interaction field determines the ultimate direction of magnetization of the particle. The particle will be magnetized in the direction of the larger of these fields regardless of its intrinsic switching field. Thus the total magnetization will be proportional to the number of particles for which the above condition holds. For a gaussian distribution of interaction field magnitudes, and assuming unidirectionally oriented particles, the anhysteretic magnetization curve is then the normal error function. On the other hand, for a random particle orientation the anhysteretic magnetization curve has lower slope and is less linear (CUMMINGS and MEE [1963]). The resulting predicted anhysteretic magnetization curves agree well with the shape of measured curves for oriented and non-oriented acicular γ Fe_2O_3 tape particles given in Chapter 6, Fig. 6.7. The essentially non-linear nature of the anhysteretic magnetization process has been demonstrated above by the dipole field model. Best linearity, however, is obtained for low applied fields thus satisfying the requirement for linear analogue recording.

A development of the above analysis could take account of the field conditions occurring in ac bias recording, although to do so rigorously would involve considerable complication.

§ 5.2. *Preisach Diagram*

A formal two-dimensional plot of particle switching fields and interaction fields will now be used to illustrate recording conditions in tapes consisting of interacting single domain particles. Such a plot is known as a Preisach diagram (PREISACH [1935]). Of course, in such a simple plot, it is not possible to take into account the vectorial addition of fields and particle magnetizations considered above. However, the simplifying assumption that the particles are oriented along the applied field direction, (x), is approximately correct. In this case the active component of the interaction field is parallel to the particle axis and to the applied field. In the diagram, the particle switching fields, H_c', and the resolved interaction fields, $H_i(x)$, are plotted as abscissa and ordinate, respectively. Each particle is represented by a point on the diagram, and the distribution of remanent magnetization may then be assessed from the density distribution of such points if the contributions of particle magnetizations to the total

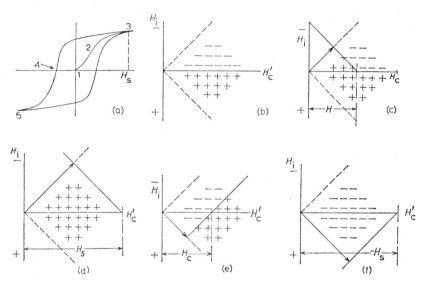

Fig. 2.17.   Preisach diagram representation of magnetization cycle:
a. Magnetization cycle assuming interaction.   d. Stage (3).
b. Demagnetized sample stage (1).   e. Stage (4).
c. Stage (2).   f. Stage (5).

magnetization is known. In a specimen demagnetized with an ac field gradually reduced to zero, all particles with positive interaction fields will be magnetized in the positive direction and those with negative interaction fields in the negative direction. Positively and negatively magnetized particles are illustrated by plus and minus signs in the Preisach diagram corresponding to the demagnetized case (Fig. 2.17b).

In keeping with the previous analysis, the H_i values are plotted with a symmetrical distribution about $H_i = 0$. A distribution of H_c' values about some finite value is also shown in accordance with practical results. On applying a magnetic field, H, those particles will be magnetized for which

$$H \geqslant H'_c + H_i. \tag{2.13}$$

Thus, magnetization will occur up to a line for which $H'_c + H_i = $ const. $= H$ in the Preisach diagram, as illustrated in Fig. 2.17c. Assuming, for now, that the intrinsic remanent magnetization is the same for all particles, then the number of particles switched positively

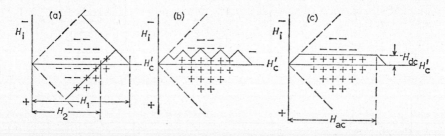

Fig. 2.18.   Preisach diagram representation of anhysteretic magnetization:
a. Magnetization after first cycle.
b. Remanent magnetization after reducing ac field to zero in five cycles.
c. Remanent magnetization after reducing ac field to zero in a large number of cycles.

in Fig. 2.17c represents the remanent magnetization from point 2 on the magnetization curve (Fig. 2.17a). Figs. 2.17d through f correspond respectively to the remanent magnetization from positive saturation, negative coercivity and negative saturation.

The anhysteretic magnetization may be described in terms of the Preisach diagram. Fig. 2.18a shows the magnetization distribution after applying $H_1 = H_{ac} + H_{dc}$ followed by $H_2 = -(H_{ac} - \frac{1}{2}\varDelta) + H_{dc}$, where $\frac{1}{2}\varDelta$ is the reduction of the ac field between successive peaks. On reducing the ac field to zero, the particles of small H_i are magnetized in the dc field direction as shown between the sawtooth boundary and the abscissa in Fig. 2.18b. If many ac cycles occur during the reduction of the ac field to zero, then the sawtooth boundary approaches a straight line parallel to the H'_c axis as shown in Fig. 2.18c. The Preisach diagram corresponding to the modified anhysteretic magnetization process is shown in Fig. 2.19. By comparison with Fig. 2.18*b* and *c* it is seen that a smaller magnetization is achieved in this modified process for a given dc field, in agreement with the experimental curve of Fig. 2.13. The conditions of the simple bias shell theory described in Chapter 2, § 2 can also be studied on this diagram, and here all particles would lie on the abscissa. The much greater recording sensitivity is easily understood for this case.

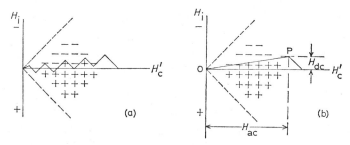

Fig. 2.19.   Preisach diagram representation of modified anhysteretic magnetization process:

a. Remanent magnetization after reducing ac field to zero in five cycles.

b. Remanent magnetization after reducing ac field to zero in a large number of cycles.

Thus the Preisach diagram presents a useful illustration of different magnetization processes in magnetic tapes and is particularly useful in showing the modified anhysteretic process. The usefulness of the diagram would be extended if it were assumed that the particle density distribution function represents the remanent magnetization distribution function and also that this function is statistically stable. The distribution has been measured by a variety of magnetization processes (DANIEL and LEVINE [1960a], WOODWARD and DELLA TORRE [1959, 1961]). The remanent magnetization may then be assessed by integration of the distribution over the areas of the diagram remagnetized by the magnetization process. However, stability of the distribution is difficult to prove for the whole diagram. In addition, the simplifications assumed for the diagram make it difficult to calculate the magnetization accurately for any practical case, although, as will be shown in the next section, approximate agreement with experimental curves is obtained at moderate magnetization levels. It is therefore concluded that the diagram retains approximate validity for qualitative description of magnetic recording conditions in materials exhibiting interparticle interactions.

The simple assumptions in this section of the particle interaction fields cannot be justified when studying the detailed behaviour of a few particles, and coercivities calculated on such assumptions are somewhat higher than those obtained for a detailed calculation (BROWN [1962]). However, it is possible that a combination of the magnetization-dependent and magnetization-independent interaction field models discussed may yield results in agreement with the experimental magnetization behaviour of tape materials (DANIEL and WOHLFARTH [1962]). For instance, in studying the phenomena of creep of magnetization level with successive applications of the same field, the experimental results may be explained by assuming that a certain percentage of the particles depend for their magnetization direction only on the average magnetization level whereas the rest depend on local field conditions (NÉEL [1959]).

In this section the important role played by particle interaction fields in magnetic tapes has been developed and illustrated with respect to the anhysteretic magnetization processes. The next step is to adapt the anhysteretic magnetization process, including interaction, to the field conditions found in magnetic recording.

§ 6. MODIFICATIONS FOR MAGNETIC RECORDING CONDITIONS

The ac bias recording process is basically described by the modified anhysteretic magnetization process. That is to say, in normal ac bias recording of long wavelengths on the tape, the maximum magnitude of the ac field is sufficient to saturate the tape and a tape element experiences enough cycles of diminishing bias field to establish true anhysteretic magnetization. However, the recording process is considerably more complicated than the simple modified anhysteretic magnetization process described in the previous section, due to the asym-

metrical field contours of a conventional recording head. It is necessary, then, to adapt the modified anhysteretic magnetization process to
the field conditions occurring in recording. Finally, further departure
from the conditions described occur for short signal wavelengths since
the ratio H_{sig}/H_{ac} can no longer be considered constant and also since
the tape no longer experiences a monotonic decrement of ac bias from
the maximum value.

Since the modified anhysteretic magnetization characteristic is

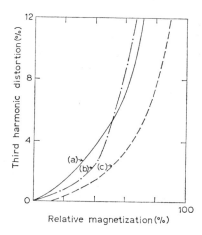

Fig. 2.20.   Magnetization linearity:
(a) Ideal anhysteretic magnetization.
(b) Modified anhysteretic magnetization.
(c) ac bias recording.

markedly dependent on the maximum value of the ac field (Fig. 2.13)
it is to be expected that the overall recording characteristic will be
represented by some suitable averaging of the modified anhysteretic
curves in accordance with the different maximum fields experienced
by layers in the tape at different distances from the recording gap.
Furthermore, it is to be expected that the recording characteristic will
then have a different sensitivity and linearity from that for the modified
anhysteretic magnetization process. The relative linearity of these
characteristics may be compared, using an ac field sufficient to give
maximum initial susceptibility; in Fig. 2.20 the linearity is represented
by the percentage of third harmonic distortion produced by these
characteristics for a sinusoidal magnetizing signal. The curves refer to
measurements made on a tape consisting of oriented acicular particles

of γFe_2O_3. The tape remanent magnetization measurements used for Fig. 2.20 were made with a search coil and fluxmeter (Ch. 6, § 3) and

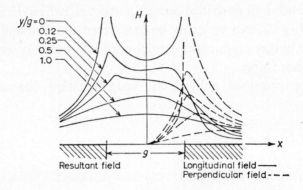

Fig. 2.21a.   Field distribution around recording gap (DUINKER [1957]).

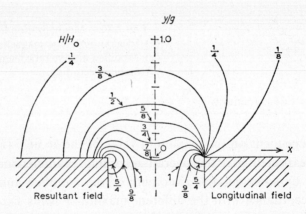

Fig. 2.21b.   Lines of equal relative field strength around recording gap
(DUINKER [1957]).
H_0 = field deep inside gap.

thus correspond to the longitudinal magnetization component only. In the case of the ac bias recording curve a small ($\approx 10\%$) perpendicular component occurring at high magnetization levels was ignored. Thus it is seen from Fig. 2.20 that the recording process is more linear than either the ideal or the modified anhysteretic magnetization processes. However, it is not possible to attribute this linearization in a simple way to the recording field configuration since such linearization occurs for a large range of recording head gap lengths. The effects of the recording field configuration on the recording transfer characteristic will be considered in more detail.

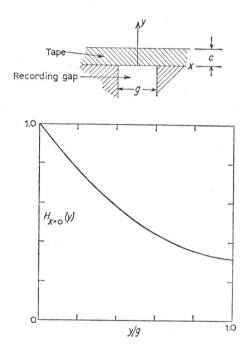

Fig. 2.22. Relative dependence of field in centre plane of recording gap on distance from gap.

§ 6.1.   *Recording Field Configuration*

The relative field amplitudes in the vicinity of the gap (length g) in a ring type of recording head are plotted in Fig. 2.21a (DUINKER [1957]). The corresponding lines of equal field strengths are shown in Fig. 2.21b, indicating the applicability of cylindrical magnetization zones for $y/g > 0.4$, but showing departures from this simple model for tape layers closer to the recording head. In general, it is seen that the resultant amplitude diminishes for increasing relative distance (y/g) from the gap: it achieves a maximum value along the gap centre plane for $y/g > 0.5$, and near the gap edge plane for $y/g < 0.5$. Examination of the components of the recording field (right hand side, Fig. 2.21a) indicates that the field at the gap centre plane is entirely longitudinal whereas at the gap edge plane it rotates through 45°. In addition, considering the longitudinal component alone, the field maximum shifts from the gap edge plane to the gap centre plane when y/g exceeds 0.12.

Tape elements in the coating will, therefore, experience different

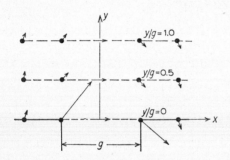

Fig. 2.23. Vector field diagram around recording gap.

maximum fields determined by their relative separation from the surface of the recording head. At the gap centre plane the calculated variation of the head field $H_{x=0}(y)$ is given by (DANIEL [1953])

$$H_{x=0}(y) = f(y/g) = k\tan^{-1}(g/2y)/\pi , \qquad (2.14)$$

where k is a constant. This function is plotted in Fig. 2.22 indicating that a significant change in recording field magnitude through the tape thickness is obtained under practical recording conditions.

As already indicated, rotation of the recording field direction occurs during the passage of a tape element past the recording gap. Concurrently, from the point of view of the recording function, the field magnitude varies in two distinct ways, which are illustrated in Fig. 2.23. For the tape element at a distance $y/g = 1.0$ the recording field rotates from $+ 45°$ to $- 45°$ at the leading and trailing edges, respectively, with a shallow maximum amplitude at the gap centre line.

Similarly, for $y/g = 0.5$ the field increases to a maximum at the gap centre line. For tape elements very close to the head, however, large field maxima are experienced at the gap edges at $\pm 45°$, respectively. Since the resolution achieved in recording and reproducing with ring type heads deteriorates rapidly with separation from the head, it is possible to describe long and short wavelength recording processes in terms of the two recording field patterns respectively described above.

§ 6.2.   *Long Wavelength Recording*

It is assumed that, to a first approximation, the longitudinal component of the recording field predominates in the recording process for long wavelength signals. Except for the surface layer, where $y/g < 0.12$, the tape experiences a maximum longitudinal field at the gap centre plane. Any effects of the subsequent rotation of the recording field as the tape leaves the centre plane will be attenuated, first, by the anisotropy of the oriented tape particles and, second, by the anisotropy of the external tape demagnetizing field. Adaptation of the modified anhysteretic magnetization characteristics to the long wavelength ac bias recording function may then be achieved by applying the field gradient function of eq. (2.14) to the modified anhysteretic remanent magnetization $I_{ar}''(H_{dc} \rightarrow 0)$. Assuming for the moment that this is given by the integration over the remagnetized zone of the Preisach diagram, Fig. 2.19b, then

$$I_{ar}''(H_{dc} \rightarrow 0) = \int_0^{H_{dc}} \int_{\left(\frac{H_{ac}}{H_{dc}}\right)H_i}^{H_{dc}+H_{ac}-H_i} I(H_i, H_c') \, dH_i \, dH_c' . \qquad (2.15)$$

In order to achieve this, the distribution function $I(H_i, H_c')$ must be determined. Some success in this direction has been achieved by assuming that the interaction field (H_i) around a particle is independent of its switching field (H_c') (CUMMINGS [1962]). In this case eq. (2.15) may be written

$$I''_{ar}(H_{dc} \to 0) = \int_0^{H_{dc}} \mathscr{I}_1(H_{dc} + H_{ac} - H_i) \cdot I_2(H_i)\, dH_i$$

$$- \int_0^{H_{dc}} \mathscr{I}_1\left(\frac{H_{ac}}{H_{dc}} H_i\right) \cdot I_2(H_i)\, dH_i, \tag{2.16}$$

where $I(H_i, H'_c) = I_1(H'_c) \cdot I_2(H_i)$, and $\int I_1(H'_c)\, dH'_c = \mathscr{I}_1$.

However, as can be seen from the Preisach diagram in Fig. 2.18c,

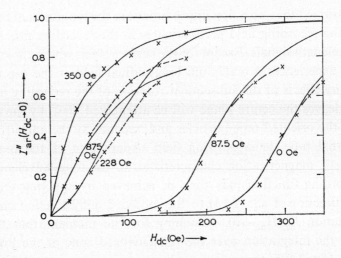

Fig. 2.24.   Computed family of modified anhysteretic magnetization curves.
$\times$ = experimental measurements.

$\mathscr{I}_1(H'_c)$ is given by the initial anhysteretic susceptibility curve and $I_2(H_i)$ by the derivative of the ideal anhysteretic magnetization curve. Graphical integration of eq. (2.16) for different values of H_{dc} is then possible yielding a family of modified anhysteretic magnetization characteristics which are compared with experimental curves in Fig. 2.24. In general, good agreement is obtained except for the higher magnetization levels where it is supposed that the simplifying assumptions no longer hold. Quantitative analysis of the initial linear portion of the recording characteristic should then be possible using this approach.

For small applied dc fields the modified initial anhysteretic susceptibility is independent of the dc field and has a pronounced dependence on the ac field magnitude. In this case eq. (2.15) may be simplified to

$$I''_{ar}(H_{dc} \to 0) = \left(\frac{H_{dc}}{H_{ac}}\right) \int_0^{H_{ac}} H'_c \cdot I(H'_c)\, dH'_c. \qquad (2.17)$$

Eq. (2.17) shows that the variation with ac field of the modified initial susceptibility, $s_\mu(H_{ac}) = I''_{ar}(H_{dc} \to 0)/H_{dc}$, is then a function of the distribution of intrinsic particle switching fields. For large ac fields, then, the initial susceptibility is inversely proportional to the ac field magnitude as indicated by the dotted curve in Fig. 2.14. Considering, now, the application of a recording head field described by $f(y/g)$ in eq. (2.14) to a tape of coating thickness c, the magnetization acquired

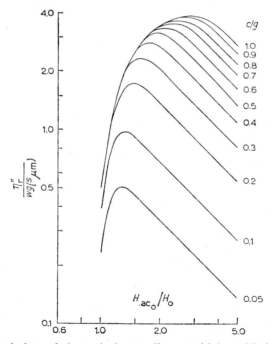

Fig. 2.25.   Variation of theoretical recording sensitivity with bias for different relative coating thicknesses, c/g (DANIEL and LEVINE [1960b]).

for a small applied signal will correspond to the integration of $s_\mu H_{dc}$ through the tape thickness or

$$I''_{ar}(H_{sig}) = \frac{g}{c} \int\limits_{0}^{c/g} I''_{ar}(H_{dc} \to 0) \, d(y/g) , \qquad (2.18)$$

and $I''_{ar}(H_{dc} \to 0) = s_\mu[H_{ac_0}, \; f(y/g)] \cdot H_{dc_0} [f(y/g)], \quad (2.19)$

where H_{dc_0} and H_{ac_0} are the field values inside the recording gap. Using an empirical relationship for $s_\mu(H_{ac})$ and eq. (2.14) for $f(y/g)$, the integration of eq. (2.18) may be carried out for different ratios of coating thickness to gap length. The resulting family of curves is given in Fig. 2.25 (DANIEL and LEVINE [1960b]), in terms of the recording sensitivity, η''_r. where

$$\eta''_r = \frac{\text{Tape flux}}{\text{Field in head gap}} = 4\pi wc \, I''_{ar}(H_{sig})/H_{dc_0} . \qquad (2.20)$$

Also, $s_{\mu m}$ and H_0 are normalizing functions for the empirical modified anhysteretic susceptibility curve. From Fig. 2.25 it is seen that, with increasing coating thickness, the recording sensitivity and the ac field required to give maximum sensitivity both increase; for very thin coatings the recording sensitivity curve approaches the modified anhysteretic susceptibility curve. It is also evident that a limit to the increase of sensitivity with thickness occurs due to the reduced sensitivity of the 'near' layers occurring when sufficient ac field is applied to magnetize the 'far' layers.

The long wavelength recording process has been described in terms of the modified anhysteretic magnetization process adjusted to account for the gradient of the recording field in the centre plane of the recording gap. Theoretical recording sensitivity curves are thereby obtained which are in good qualitative agreement with practical recording behaviour. However, the recording process appears to contain some further loss which is not accounted for by the approach described. This loss may possibly be ascribed to a demagnetizing effect occurring during the anhysteretic magnetization process and to the perpendicular field component occurring especially in the near layers (DANIEL and LEVINE [1960b]). The near layer behaviour is accentuated in short wavelength recording conditions due to the wavelength dependent exponential separation loss occurring on reproduction (see Ch. 4).

§ 6.3.   Short Wavelength Recording

If it is assumed that the above theory is applicable to short wavelength recording with ac bias, it is to be expected that, since the near layers contribute most of the tape surface flux due to the recording, the recording sensitivity curves will approach the lower curves in Fig. 2.25. In practice, the bias for maximum sensitivity does decrease as predicted. However, the predicted inverse proportionality of sensitivity and bias amplitude for large bias amplitudes is not obtained: a consequence of this prediction is that the magnetization is constant at all depths in the coating in the overbiassed condition. In practice a reduced magnetization occurs in the surface layers which may be attributed to the $\pm 45°$ direction of the maximum recording field depicted in Fig. 2.23. For large bias amplitudes, when the field at the gap centre line is sufficient to saturate the tape, recording takes place entirely beyond the trailing edge. Thus a modified anhysteretic magnetization process takes place with the maximum field at $45°$ to the particle orientation, rotating to $90°$ as the field decreases to zero. For small bias amplitudes a further complication arises since irreversible magnetization occurs at both gap edges and interference effects may be added to the process (see Ch. 3 § 1.1). This latter effect, however, does not occur under conventional ac bias recording conditions.

A further possible loss occurring at high frequencies is due to the departure from the anhysteretic magnetization condition that the applied dc or signal field be constant during the magnetizing process. If the amplitude of the signal changes during the decrement of the bias, then the boundary line of the remagnetized zone on the Preisach diagram (Fig. 2.19b line OP) is no longer straight. At high frequencies this effect can lead to a recording demagnetization effect in which a reduced maximum remanent magnetization is achieved for a lower signal field than that giving a maximum low frequency magnetization (SCHWANTKE

[1961]). The corresponding Preisach diagram is shown in Fig. 2.26, where the signal changes by one complete cycle during the rundown of the bias through the irreversible magnetization process range of the recording medium. The signal amplitude is large enough to produce a maximum slope (θ) greater than 45° for the remagnetized boundary zone, thus causing a recording demagnetization loss. Maximum output at short wavelengths is obtained for a signal level corresponding to $\theta = 45°$. This loss is minimized by reducing the distribution of particle switching fields in the tape and also by increasing the field decrement at the trailing edge of the recording head gap. The latter function can

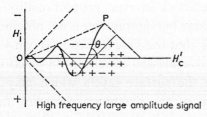

High frequency large amplitude signal

Fig. 2.26.   Preisach diagram representation of recording demagnetization at high frequencies.

be optimized by adjusting the bias amplitude so that optimum recording sensitivity of the surface layers occurs with the maximum field decrement. The additional losses occurring due to overbiassing may be partially attributed to the lower field decrement occurring at the point of recording in the surface layers.

A further loss may occur at very short wavelengths due to the change of the longitudinal position of the point of recording with depth into the coating. As the first recording model indicated in an exaggerated form, (see § 2), the recording zone is not perpendicular to the direction of motion of the tape, and different layers will be recorded in different positions with possible destructive interference at short wavelengths. This loss will, of course, be somewhat attenuated by the increased separation loss at short wavelengths.

Finally, when the tape moves away from the recording head, and is no longer in contact with the high permeability pole pieces, the shunting action of the pole pieces is removed and the demagnetizing field inside the tape increases. Suffice it to say, the demagnetization loss increases for thick tapes and short wavelengths. Depending on its magnitude, the demagnetizing field will produce a reversible and an irreversible reduction of magnetization. This is illustrated in Fig. 2.27 in which the model hysteresis loop of Fig. 2.11 is used with account

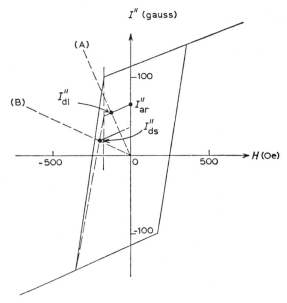

Fig. 2.27.   Model hysteresis loop for magnetic tape showing self demagnetization losses for
(A) Long wavelength recording.
(B) Short wavelength recording.

taken of the reversible susceptibility, which is about $1/4\pi$ for the oriented γ Fe_2O_3 tape loops shown in Fig. 2.10a. The demagnetization lines for long and short wavelengths are drawn as dashed lines (A) and (B) respectively. For a long wavelength recording of magnitude I''_{ar} a small reversible loss to I''_{dl} is experienced. Some of this lost magnetization will be recovered on reproduction since the demagnetizing field is reduced particularly at the tape surface in contact with the reproducing head. On the other hand, for a short wavelength recording of similar magnitude, I''_{ar}, a reversible loss and an irreversible loss occur reducing the magnetization to I''_{ds}, as shown in the diagram. In this case, on reproduction, only a small amount of reversible magnetization is recovered. Although such irreversible loss appears to limit severely the magnetization recordable with linearity at short wavelengths, oriented acicular γ Fe$_2$O$_3$ particle tapes appear to have a sufficiently

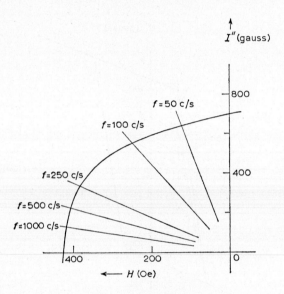

Fig. 2.28.   Demagnetization lines for sinusoidally magnetized Vicalloy tape for different frequencies f.
Tape thickness = 0.003 inches.
Tape speed = 15 inch/sec (LOGIE [1953]).

high ratio of coercive force to maximum remanent magnetization for this loss to be relatively small. Of course, other materials with high ratio of magnetization to coercivity, such as Vicalloy (see Ch. 6), would suffer more severe losses from irreversible self demagnetization. The demagnetization coefficients for a range of sinusoidally recorded frequencies for a relatively thick Vicalloy tape are shown in Fig. 2.28 (LOGIE [1953]), indicating severe self demagnetization losses for these relatively unfavourable circumstances. Materials with high reversible susceptibilities will suffer larger reversible demagnetization losses and this characteristic may materially affect the short wavelength response.

CHAPTER 3

MAGNETIC RECORDING PROCESS –
ZERO AND UNIDIRECTIONAL BIAS

§ 1. ZERO BIAS

In recent years, direct recording without bias has been widely used. There are two major applications of this type of recording. Firstly, for data recording where two or three signal magnitudes only are to be stored, the three significant magnetization states of the medium are used: that is, positive and negative saturation, and zero magnetization. These magnetization conditions are usually obtained directly from the signal field of the recording head. The second important area for direct recording is in the storage of a frequency modulated signal. In this case, amplitude linearity is unimportant and the linearity of the system depends on the speed constancy of the tape. From the recording point of view, the important characteristics are high resolution and large signal-to-noise ratio. Conditions leading to high recorded levels and high resolution will be discussed after a description of the basic zero bias recording phenomenon using conventional ring heads for longitudinal recording.

§ 1.1. *Sine Wave Recording*

Long Wavelengths

Using the model (see Ch. 2) of the cylindrical magnetization zone whose diameter is proportional to the signal, described initially for the ac bias recording process, the process of recording can be described in geometrical terms by considering the growth and decay of the zones, following the signal variations, and the corresponding magnetization pattern imprinted on a magnetic coating moving through the zones. It will be recalled that this model is applicable to the case where both the coating thickness and the recorded wavelength are large compared to the recording head gap length and where the anisotropies of the tape

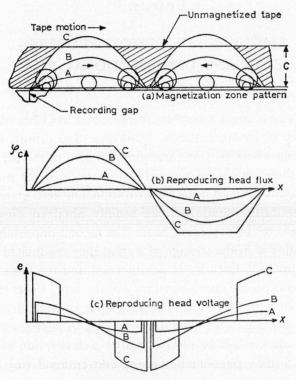

Fig. 3.1.   Long wavelength recording without bias.

favour magnetization due to the longitudinal component of the recording field only.

The simplest recording condition occurs when a sinusoidal signal, whose amplitude variation is small during the passage of a tape element across the recording zone, is applied to the recording head without bias. The magnetization pattern imprinted on the tape for a sinusoidal signal is shown in Fig. 3.1a in which the arcs indicate the course of the cylindrical magnetization zone diameters. The three

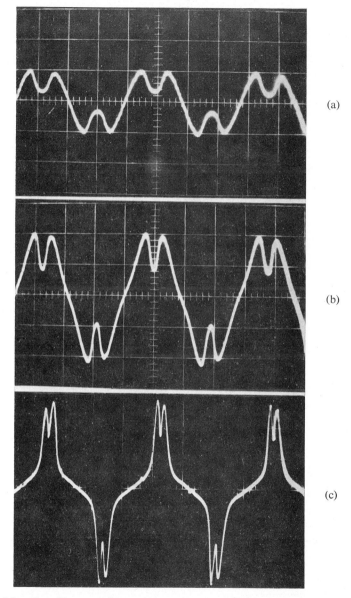

Fig. 3.2. Zero bias recording; a–c increasing sinusoidal recording signal.

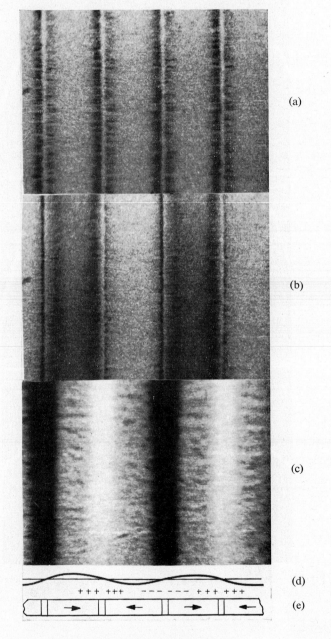

Fig. 3.5.　Sinusoidal recording. Zero bias.
a. Zero field.　　　b. Horizontal field.　　　c. Vertical field.
d. Recording signal.　　e. Tape magnetization.

boundaries shown in Fig. 3.1a correspond to signal magnetization zones whose maximum diameters are less than, equal to, and greater than the coating thickness for Curves A, B and C, respectively. Thus, direct recording of a sinusoidal signal whose wavelength on the tape is long compared to the coating thickness is essentially variable depth magnetization following the signal amplitude until the magnetization

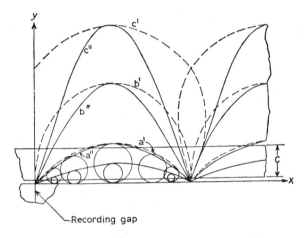

Fig. 3.3.   Recording zone contours. Zero bias.
Dotted line – Zone boundary for sinusoidal signals.
Solid line – Boundary of cylinder diameter $y = R_c$.
c – Tape coating thickness.

zone diameter becomes equal to the tape thickness (Curve B, Fig. 3.1a). For larger signals, the recording field penetrates beyond the remote side of the tape coating, and further magnetization is obtained by the recorded zones becoming steeper sided and pushing into the unmagnetized tape area indicated in Fig. 3.1a.

Detection of a zero bias recording with a reproducing head will produce a flux waveform in the head as shown in Fig. 3.1b corresponding to the three recording levels depicted in Fig. 3.1a; the corresponding voltages are shown in Fig. 3.1c. It can be seen that, for long wavelengths, the distortions predicted are similar to those found in practice (Fig. 3.2a, b, c).

The recorded area becomes modified when the diameter of the cylindrical zones approaches half the wavelength of the recorded signal as shown in Fig. 3.3. The solid lines are sinusoidal curves representing the instantaneous magnitude of the recording field. Thus, for curves a″, b″, and c″

$$y = R_c = R_{c(max)} \sin (2\pi x/\lambda) . \tag{3.1}$$

The larger areas swept out by the cylindrical zones are shown by the dotted curves a′, b′, and c′. As the magnitude of the signal is increased, a condition is reached where the boundaries of the recorded zones are perpendicular to the tape surface where the recorded magnetization passes through zero (curve b′). In this case, the change of direction of magnetization, when the recording signal passes through zero, takes place at a maximum rate. For this condition

$$(dy/dx)_{x \to 0} = 2 = 2\pi R_{c(max)}/\lambda . \tag{3.2}$$

Thus,

$$R_{c(max)} = \lambda/\pi , \tag{3.3}$$

where λ is the wavelength of the recording on the tape. Larger signals (curve c′) produce overlapping magnetization zones for successive half wavelengths. This condition produces a change of phase of the recorded magnetization which can be significant in short wavelength recording such as FM recording.

Practical Recording Gaps

In practice, the recording gap length is often comparable with the tape coating thickness and the magnetization zone departs from the simple cylindrical zone considered above. For a finite recording gap length it has been shown in Chapter 2, § 6.1 that in the gap centre plane the recording field is longitudinal, reaches its maximum value, and decays with separation from the head as shown in Fig. 2.22. Thus the maximum field at the surface of the head is finite rather than infinite as given by the inverse separation function of an infinitely small gap.

Assuming that the remanent magnetization achieved by an element of tape in its passage through the recording field zone is characterized by the maximum field occurring at the gap centre plane, the recording characteristic for a finite gap recording head can be approximately calculated as follows. The remanent magnetization characteristic for oriented acicular particle tapes may be closely approached for instance by the empirical relation

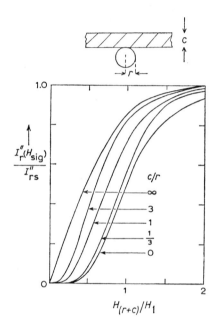

Fig. 3.4. Empirical zero-bias recording characteristics for different tape coating thicknesses c, recorded with a single wire of radius r.

where
$$I''_r(H_{sig}) = I''_{rs}\left\{ (H^4_{sig})/(H^4_1 + H^4_{sig}) \right\}, \qquad (3.4)$$
$$I''_r(H_{sig})/I_{rs} = 0.5 \text{ for } H = H_1 .$$

The field in the gap centre plane may be closely approximated by that of a wire of radius r in contact with the coating surface; thus

$$H_{x=0}(y) = 2i/(r + y) . \qquad (3.5)$$

Substitution in eq. (3.4) gives

$$I''_r(i, y) = I''_{rs}\left/\left[1 + \left\{ \frac{H_1(r + y)}{2i} \right\}^4 \right]\right. . \qquad (3.6)$$

The tape magnetization may then be calculated by integrating eq. (3.6) over the tape thickness, c, resulting in the curves of Fig. 3.4 for various values of c/r. It can be seen that linearity increases with decreasing radius; in the limiting case $c/r \to \infty$, exact proportionality is obtained from zero applied field to a high magnetization level, as predicted by the simple model.

The non-linearity of the zero bias recording process may be illustrated experimentally by examining the recorded tape surface flux pattern with a colloidal magnetite indicator. A sinusoidal signal recording may be made visible showing clearly the finite tape lengths which are unmagnetized as the signal passes through zero (Fig. 3.5). That the recorded pattern is as depicted in Figs. 3.5e and 3.1a, is illustrated by the colloidal magnetite pattern on the same recording, made in the presence of a horizontal and vertical field (Figs. 3.5b and c). The function of these fields is to polarize the magnetite particles in the ink to produce selective attraction or repulsion, depending on the relative polarity of the surface induction (see e.g. MEE [1950]).

The application of the simple model to sinusoidal signal recording without bias gives a qualitative picture of the distortions occurring due to the growth of the cylindrical magnetization zones through the coating thickness and to their mutual interference. The latter effect has been shown to be due to a shift of the effective point of recording on the cylindrical zone surface, depending on the rate of growth or decay of the magnetization volume with respect to the motion of the tape. The resulting non-sinusoidal magnetization pattern is shown in Fig. 3.3. Similar but more complicated processes occur when recording with finite gap lengths. The partial linearization of the long wavelength recording characteristic when recording with gaps of various lengths has been described, and it now remains to consider the effect of finite gap lengths on the recording process for different wavelength ranges. It has been shown (STEIN [1961]), that the recording process is essentially a function of the wavelength to gap length ratio (λ/g) for a given element of tape. When this is large, distortion occurs in an analogous fashion to that described for an infinitely small gap due to the shifting of the recording point during the magnetization cycle. In this case, the recording point moves from the trailing to the leading edge of the gap during each half cycle. Hence, the parts of the recorded signal near the signal zeros will be recorded twice, since the recording point jumps from the leading to the trailing edge as the signal passes through zero.

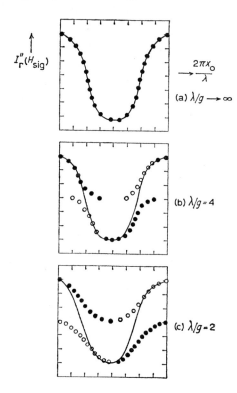

Fig. 3.6. Magnetization waveforms for sine wave recording without bias using finite gap (STEIN [1961]).
・・・・・ First recording field.
∘∘∘∘∘ Second recording field.
———— Resultant magnetization.

Medium Wavelengths

As the wavelength of the recorded signal is reduced and becomes comparable with the gap length, the mutual interference effects, due to tape elements being subjected to fields of opposite polarity during their passage across the recording gap, become more significant and lead to increasing distortion of the recorded waveform. This is illustrated in Fig. 3.6 in which the first and second magnetization fields are indicated in terms of relative distance along the tape, for relative recorded wavelenghts $\lambda/g = \infty$, 4 and 2. For simplicity it is assumed that the gap field function is a cosine function about the gap centre line and that the tape magnetization is proportional to the applied field. The inherent distortion of this recording process is clearly seen to increase as λ/g approaches unity. Taking account of a non-linear magnetization characteristic and the variation of the field function and

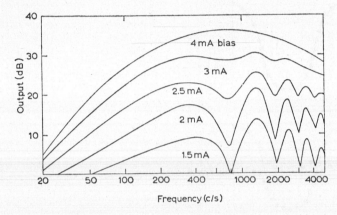

Fig. 3.7.   Interference of recorded signals for various ac bias levels.
Tape: oriented acicular particles. Record gap length: 1.73 mil.
Tape speed: $1\frac{7}{8}$ inch/sec.

amplitude for different tape layers produces further modifications to the inherent distortion.

A further important interference effect occurs in the medium wavelength range when a relatively thin tape coating is recorded in contact with the recording head. If $c/g < 0.5$ then all layers in the tape experience two field maxima at the gap edges (Fig. 2.21a) and double recording takes place due to these two maximum fields which are directed at 90° to each other. It is then found that the resultant magnetization undulates periodically with recorded wavelength. The minima in the recorded magnetization occur at wavelengths which are related to the recording head gap length by

$$g = (n - 0.25)\,\lambda; \quad n = 1, 2, 3\,. \tag{3.7}$$

Typical frequency response curves are shown in Fig. 3.7 for oriented acicular particles. The families of curves shown have been recorded with different levels of ac bias, and it may be noted that the effect of the bias on the undulations is solely to attenuate them, their position and number remaining unchanged.

It has been demonstrated (DANIEL [1960]), that for successive equal orthogonal fields, the attenuation of the first magnetization by the second is less than 50% for magnetization levels up to 80% of saturation by the second field. Thus, in this magnetization range it can be assumed that the final tape magnetization will be the superposition of two orthogonal magnetization components. Similar anlayses (DANIEL [1960]), WOODWARD and PRADERVAND [1961]), of the superposition of two non-unidirectional magnetization waveforms confirm the experimental relationship of eq. (3.7). The positions of the interference minima with respect to the recorded wavelength and recording head gap length will depend on the ratio of longitudinal to perpendicular magnetization. Eq. (3.7) refers to equal longitudinal and perpendicular components. As the latter is decreased, the minima occur at longer wavelengths until, for $I''_{r(y)} = 0$,

$$g = (n - 0.5)\,\lambda \,. \tag{3.8}$$

The above analysis yields results in agreement with experiment. Superposition of the two magnetizations can only be explained if some satisfactory reason can be advanced why the second field should magnetize a different set of particles to the first. In the case of acicular particles, the simple Stoner-Wohlfarth magnetization model predicts minimum switching fields at 45° to the applied field. Thus, for two small orthogonal fields, different sets of particles are involved in the two acquired magnetizations. However, when the two fields are more nearly equal and opposite, the mechanism is more likely to be due to a redistribution of local internal fields after the first magnetization, leading to different particles having the lowest switching fields for the second applied field. Of course, for large magnetization levels, magnetization by a second field can only be achieved at the expense of the first magnetization. This is shown clearly in Fig. 3.8 for superposed pulses where a large initial magnetization (Fig. 3.8*a*) is succeeded by increasing amplitudes of reverse magnetization (Figs. 3.8*b–d*). Fig. 3.8*e* shows the reverse magnetization zone alone. The corresponding repro-

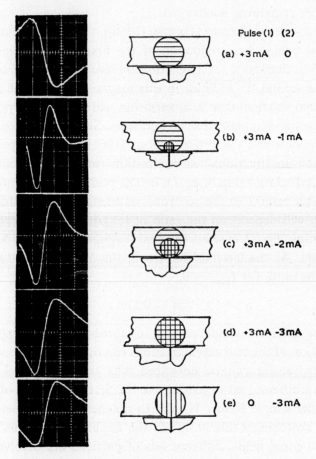

Fig. 3.8.   Superposition of two recording pulses.

ducing head voltage waveforms shown on the lefthand side of the picture indicate the high resolution of the small zone of reverse field (Fig. 3.8*b*). However, the reverse magnetization spreads with increasing reverse field until, for equal and opposite successive fields, the final magnetization appears to be similar to that of the second field alone (Figs. 3.8d and e).

Short Wavelengths

As the recorded wavelength is reduced and becomes less than the gap length, each element of tape becomes subjected to a number of signal field maxima during its passage across the gap. Interference takes place between the corresponding magnetization to such an extent that it is only the final magnetizing fields, as the amplitude is being reduced on the trailing side of the gap, that have significance in determining the final magnetization amplitude. In this range the rate of extinction of the recording field influences the shortest recordable wavelength. If it is attempted to record beyond this limit no net tape magnetization occurs and the situation is analogous to the application of ac bias without a signal.

The increase of self erasure taking place with wavelength reduction below $\lambda/g = 1$ causes the effective point of recording to move from the gap centre line towards the trailing edge. Since the recording field function varies with distance from the head there will be a further variation in the shift of the recording point from the gap centre line as a function of separation from the head. In addition, an overall attenuation of the recorded magnetization takes place due to partial erasure during the extinction of the field at the trailing edge. This attenuation is a function of the rate of extinction and occurs at longer wavelengths for tape elements separated from the head. Both phase shift and attenuation then occur in different amounts for different tape layers and contribute to the overall short wavelength losses. The ultimate short wavelength recording limit will be determined by the maximum slope of the trailing edge field for the near layers of the tape.

§ 1.2. *Pulse Recording*

NRZ Recording

To a large extent, the interference, distortion, and phase shift effects described for sine wave recording apply also to non-return-to-zero

(NRZ) pulse recording. In this system of recording, commonly used for storage of binary information, the recording field is held at a constant amplitude which is at all times sufficient, at least, to saturate the entire tape coating in the vicinity of the gap. Information is then imparted to the recording signal by a coded sequence of reversals of the signal polarity.

The choice of the optimum recording field amplitude is governed by the dependence of the recording resolution and the incidence of drop-outs at different recording levels. A dropout may be defined as an imperfectly recorded pulse which cannot be satisfactorily detected in reproduction. A dropout may occur due to a reduction in amplitude of the recorded magnetization incurred, for instance, by a non-uniformity in the coating surface, causing a local separation of the tape from the recording head as illustrated in Fig. 6.22 (Ch. 6). Also, in NRZ recording, the longitudinal position of the recorded pulse depends on the head to tape separation, and dropouts may occur due to timing errors between pulses recorded on parallel tracks. The first type of dropout may be minimized by increasing the recording field amplitude to cause saturation of all layers in the tape; however, this condition leads to poor resolution. The time displacement of the recorded pulses becomes the more serious problem in high resolution pulse recording systems. It is found that an intermediate recording field amplitude leads to minimum pulse shifting of this type and hence to an optimum dropout immunity. In practice, relatively wide record-ing gaps are used for NRZ recording to minimize the field decrement through the coating thickness and thus to improve further the immunity to recorded dropouts. Recording resolution is not badly impaired for wide gaps since the increased spread of the recording field in the direction of tape motion is offset by a relative increase of recording field decrement.

The dependence of the recording resolution and the form of the tape magnetization transition, for an instantaneous reversal of the record-ing current, will now be considered as a function of the recording field strength. For high resolution it is desirable that the trailing field of the recording head has a maximum rate of extinction as the amplitude falls through the particle switching field range. It will be assumed, as a first approximation, that the longitudinal component of the recording field is the only active part of the total field in producing tape magnetization. The dependence of the tape magnetization, after previous saturation, on the magnitude of the reverse field, may be obtained

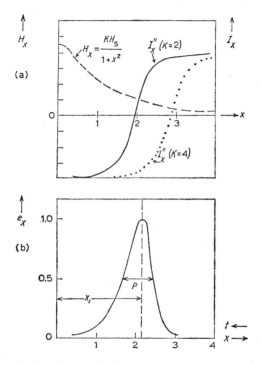

Fig. 3.9.   Recorded transition and ideal reproduction for NRZ step function
(Teer [1961]):
a. Recording field H_x, and magnetization transition I_x'' for field reversal.
b. Reproduced pulse shape e_x by differentiation of I_x''.

from the measured family of magnetization loops such as is shown in Fig. 2.10a. The longitudinal component of the recording field may be obtained from Fig. 2.21a. Assuming a value for the relative spacing from the head $y = 0.5g$, the simple formula for an infinitely small gap is approximately applicable. With these applied field and tape magnetization characteristics, the magnetization along the tape may be computed for different maximum applied field levels, assuming the coating has negligible thickness. The resulting form of the magnetization transition on a recorded tape, for an applied field reversal from $H_{max} = - KH_s$ to $H_{max} = + KH_s$, is shown in Fig. 3.9a (TEER [1961]). The shape and relative position of the magnetization reversal will depend on the tape magnetization characteristic and on the applied field shape and magnitude. For the cases illustrated, $K = 2$ and 4, it is seen that the magnetization reversal is not symmetrical: the reproduction pulse shape would be the derivative of this reversal, Fig. 3.9b, if no further distortion is introduced in the reproducing process. It may

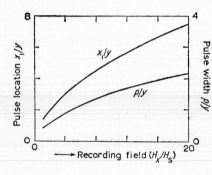

Fig. 3.10. Pulse width p, and location x_1 as a function of recording current. y-head-to-tape distance (TEER [1961]).

be noted, however, that similar reproduction pulse distortion can be attributed to the perpendicular component of the recorded magnetization, although the distortion of the longitudinal recording is probably the dominant effect. In the above diagram it is assumed that the tape is moving from left to right and hence, for the reproduced pulse, the opposite direction will correspond to increasing time. Under these conditions the reproduced pulse will therefore have a steep rise followed by a shallow decrement. Changes in the reproduced pulse shape occur for different recording field amplitudes; as the amplitude is increased, both the pulse width, p, (at 50% of maximum) and its location, x_1, change as shown in Fig. 3.10. The change of the magnetization transition on doubling the applied field ($K = 4$) is also shown in Fig. 3.9, indicating the shift of the transition away from the recording gap and the greater spread of the transition zone. This shift of the transition zone can give rise to dropouts as described earlier.

In practice, the above characteristics for pulse recording are somewhat modified when a finite tape thickness is considered along with
finite recording and reproducing head gap lengths. Starting with the
reproducing function, the reproducing head flux may be approximately
expressed as a convolution integral of the longitudinal field function of
the head (H_{xr}) and the longitudinal tape magnetization I_x''. For a tape
of thickness, c, spaced a distance, a, from the head, the core flux φ_c is
given by

$$\varphi_c = K \int_a^{a+c} dy \int_{-\infty}^{+\infty} H_{xr} I_x'' \, dx , \qquad (3.9)$$

where K is a constant.

The corresponding reproducing coil voltage is given by

$$e = K' \int_a^{a+c} dy \int_{-\infty}^{+\infty} H_{xr} \frac{\partial I_x''}{\partial x_1} \, dx , \qquad (3.10)$$

where N is the number of turns on the head coil, v is the tape velocity,
K' is a constant and $x_1 = vt$. If x_t is a longitudinal coordinate with
respect to the tape, then

$$x_t = x - vt . \qquad (3.11)$$

Eq. (3.10) may be expanded in terms of the recording head field H_{xw}
and x_t (CHAPMAN [1963]). Thus

$$e = K' \int_a^{a+c} dy \int_{-\infty}^{+\infty} H_{xr} \frac{\partial I_x''}{\partial H_{xw}} \frac{\partial H_{xw}}{\partial x_t} \frac{\partial x_t}{\partial x_1} \, dx . \qquad (3.12)$$

Here, I_x'' and H_{xw} are expressed as functions of position in the tape
(x_t, y). Thus the reproduced voltage is determined by the averaged
product of the reproducing head field function, H_{xr}, the slope of the
remanent magnetization curve, ($\partial I_x''/\partial H_{xw}$), and the slope of the
recording head field function ($\partial H_{xw}/\partial x_t$).

Optimum recording resolution will thus be obtained if the maxima of
the remanent magnetization curve slope and of the recording head
field coincide. These functions are plotted in Fig. 3.11a for a finite

recording head gap length and for a relative spacing from the head of $y = 0.57g$. Their product is the derivative of the tape magnetization with respect to distance along the tape and is therefore similar to Fig. 3.9. However, as shown in Fig. 3.11b, this function spreads considerably with increased depth in the coating, illustrating the increased average magnetization transition zone length for the whole

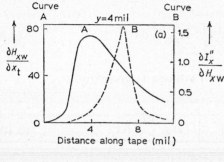

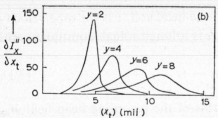

Fig. 3.11. Recorded transitions for NRZ step function.
Recording head gap length = 7 mil.
a. Recording head trailing edge field slope $\partial H_{\bar{x}w}/\partial x_t$ and magnetization curve slope $\partial I''_x/\partial H_{\bar{x}w}$ as a function of distance along the tape x_t.
b. Recorded magnetization slopes for head-to-tape spacings $y = 2, 4, 6, 8$ mil.

coating. Furthermore, on taking account of the reproducing head field function in eq. (3.12), the reproduced pulse width will be somewhat more increased. Nevertheless, it is found that measured pulse widths are even wider than predicted by the above analysis, and it is necessary to consider self-demagnetization effects in the magnetization transition region in order to account for this.

Thin coatings can therefore be seen to be advantageous for pulse recording. If a high remanent magnetization material is used to offset the reduction of coating thickness, it would be necessary to increase the coercivity correspondingly and to maintain a rectangular demagnetization characteristic in order to minimize demagnetization losses. In fact, it has been shown that, even for the ratios of remanent magnetization to coercive force found in iron oxide tapes, the demagnetization effects cause significant increases of the magnetization transition zone (TEER [1961]). Improved pulse recording resolution may therefore be expected from higher coercivity and squarer loop tape materials. Further improvement may also be expected from recording head designs which produce more rapid longitudinal decrements of the field amplitude for all layers in the tape.

RZ Recording

In this recording technique, a tape-saturating recording current of short duration is applied to the head to represent an element of information, and the tape is left unmagnetized in the absence of information. This recording system is known as return-to-zero recording and is quite distinct from NRZ recording. The magnetization zone due to unidirectional field of short duration is recorded at the leading edge of the gap as the field increases and at the trailing edge as it decreases. Hence the recording resolution is determined by the gap length. Since the recorded pulse is due to the applied field at both gap edges, there is no phase displacement at a dropout, only loss of resolution and attenuation of the magnetization. The attenuation is high, however, since the recorded pulse has no long wavelength components. Also, the recording resolution decreases much more rapidly for large signal levels than for NRZ recording, although, for fields below those required for tape saturation, the resolutions of the two methods are similar.

§ 2. UNIDIRECTIONAL BIAS

§ 2.1. *DC Bias Recording Mechanism*

In this section the magnetic recording mechanism will be considered for systems in which linearization of the transfer characteristic is obtained by the application of a unidirectional magnetic field along with the signal field. Recording with dc bias has been superseded by ac bias in many applications because of the lower background noise level obtained with the latter, and the less critical bias amplitude conditions for low distortion. However, some applications of magnetic recording are still best served by unidirectional bias systems. For instance, the efficiency of dc bias production could be advantageous in recording very high frequency signals, since the frequency losses for an even higher frequency ac bias will be large.

The function of any bias method is to produce magnetization of the tape in such a form that it is not reproduced in playback; it should, however, produce a state of magnetization which changes proportionately with any superimposed signal current applied to the recording head. In this way the non-linearity for small signal fields is avoided. As has been explained already, in the previous section, the non-linear magnetization characteristic of tape is partially offset in longitudinal recording by the non-uniform field of a narrow gap recording head (Fig. 3.4). This results in smaller relative fields for commencement of magnetization than in the case of a uniform applied field. If this small starting field is supplied by a dc bias, then unidirectional signal fields of the same polarity would be recorded without distortion. Recording of alternating signals is, however, more important. To achieve this with dc bias, a larger bias is required sufficient to produce a magnetization zone of approximately half the coating thickness.

Due to the linearizing action of small recording gaps it is preferable to maintain a large ratio of coating thickness to gap length for dc bias recording since the distortion for optimum dc bias is thereby reduced. For this reason, the simple recording model of a cylindrical magnetization zone is a good approximation to the practical conditions and will be used in the description of the process (BAUER and MEE [1961]). For a perfect tape, the criterion of non-reproducibility of the bias recording would be met since the resulting tape surface induction is non-variant. In practice, tape inhomogeneities cause local variations from the saturation magnetization level, giving rise to noise on reproduction. Lower noise levels are obtained from bias systems which produce a more random magnetization.

On adding signal to the dc bias, a depth modulation results which corresponds to the signal field. Thus, magnetic poles will be generated along the boundary with a density proportional to the rate of change of the signal. Fig. 3.12a shows the depth modulation for a constant ac signal and dc bias increasing from zero in five steps to a value sufficient to push the signal modulation through to the remote side of the coating. Fig. 3.12b shows the corresponding longitudinal tape flux pattern, and Fig. 3.12c the reproducing head voltage, assuming that

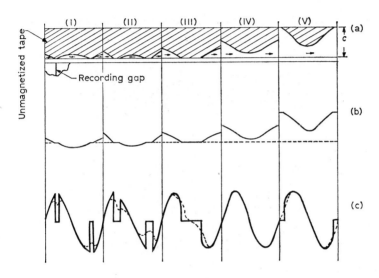

Fig. 3.12.   Dc bias recording model.
a. Tape magnetization pattern.
b. Tape flux amplitude.
c. Reproducing head voltage.

this is the derivative of the tape flux amplitude. The dotted lines in Fig. 3.12c indicate the waveforms found in practice. Examples of the corresponding reproducing head waveforms are shown in Fig. 3.13a–e.

If the tape was previously saturated in the opposite direction to the bias field direction, then the pole density and reproducing head output voltage are doubled compared with the previous case. Since the tape is premagnetized, it is not possible to obtain magnetization patterns of the type 3.12a (i) and (ii), which contain two magnetization directions. Instead, the zero bias pattern is similar to Fig. 3.12a (iii). Experimental reproducing head waveforms for increasing dc bias on premagnetized tape are shown in Fig. 3.13 *f–i*.

It can be seen from Fig. 3.12*a* that the maximum undistorted signal may be recorded when the bias level is adjusted for the bias magnetization zone to be half the coating thickness as in 3.12*a* (iv). As would

Unmagnetized tape

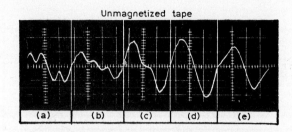

Premagnetized tape (Gain reduced)

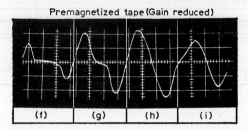

Fig. 3.13.   Reproducing head output waveforms, dc bias recording.

be expected from the model, the distortion is much more evident for large signal levels. When short wavelengths are recorded with dc bias, the separation between the magnetization boundary and the surface of the tape causes increased loss of reproducing head flux. This may be retrieved by reducing the dc bias and bringing the boundary nearer to the tape surface with consequent loss of undistorted dynamic range.

§ 2.2.   *DC Bias Recording Sensitivity*

In Chapter 2, § 1 it was pointed out that the remanent magnetization dependence on applied signal field is considerably linearized if the field is applied non-uniformly as in the case of a recording head. For ac bias, this linearizing action contributes little to the transfer characteristic since the anhysteretic magnetization process is the dominant effect. For dc bias, however, tape layers at different depths in the coating have different susceptibilities due to the change in magnitude of dc bias

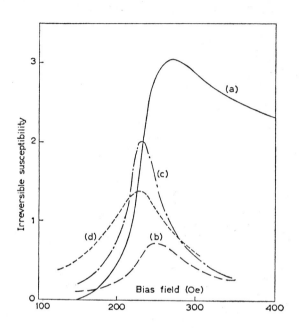

Fig. 3.14.   Initial irreversible susceptibility *vs.* bias field strength.
(a) ac bias $H_{ac}/H_{dc} = $ constant
(b) dc bias   ⎫
(c) dc bias-negative premagnetization   ⎬ (DANIEL and LEVINE [1960a])
(d) Typical recording sensitivity   ⎭

through the magnetic layer, and the resultant susceptibility is less critically dependent on bias amplitude than in the case of a uniformly applied field.

The irreversible susceptibility for a uniform applied field has a lower maximum value for dc bias than for ac bias. The dependence of susceptibility on bias amplitude for ac, dc, and dc with premagnetization is shown in Fig. 3.14*a*, *b*, *c* respectively (DANIEL and LEVINE [1960*a*]). The different shapes of the ac and dc bias curves and their different maximum amplitudes are determined by the switching field distribution. The different modes of magnetization acquisition for the three cases may be understood by reference to the Preisach diagram (Fig. 3.15). The magnetization zones are shaded on the diagram for a small constant signal field H_{dc} and three bias amplitudes. In addition,

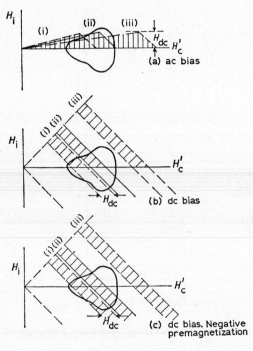

Fig. 3.15. Preisach diagrams showing magnetized zones for small constant signal (shaded area) using different bias techniques:
Bias amplitudes: (i)   0.75 H_{max}
            (ii) H_{max}
            (iii) 1.5 H_{max}

a pear shaped $I(H_i, H'_c)$ density contour line is shown on each diagram. For each type of bias, maximum sensitivity occurs when the bias is sufficient to cause maximum area change, within the $I(H_i, H'_c)$ contour, for a small change in signal. This condition corresponds to bands (ii). Also, for each type of bias, a 25 % reduction from optimum bias amplitude causes a rapid decrease in area within this contour. On the other hand, over-biassing by 50 % causes similar rapid decrease for the dc biassing techniques (Fig. 3.15b, c), but only a small loss for ac bias. The relative maximum sensitivities are also demonstrated by the

Preisach diagrams, indicating that the ac bias method magnetizes all the particles with low H_i (regardless of H_c') whereas the dc method magnetizes those for which

$$H_{dc} \geqslant H_c' + H_i , \qquad (3.13)$$

where $(H_c' + H_i)$ is measured from the 45° line corresponding to bias only in Fig. 3.15b, c.

The irreversible susceptibility curves give an adqeuate description of the recording sensitivity for long wavelength signals with ac bias, since the magnetization acquired by different layers in the coating depends largely on the ratio H_{sig}/H_{ac}, which is constant. For dc bias, however, the recording sensitivity curve is the summation of the susceptibilities of elemental layers of the tape coating, each of which has been subjected to different dc bias magnitudes determined by the recording head field function through the tape coating. In this way, the maximum dc recording sensitivity is further reduced as shown in curve d, Fig. 3.14. For wide gap recording heads, the smaller field decrement through the coating would reduce the departure from the irreversible susceptibility curve.

§ 2.3. *Short Wavelength Recording with DC Bias*

The long wavelength recording process using dc bias is reasonably distortion-free, providing the signal amplitude is small enough to limit the excursions of the cylindrical recording zone within the coating thickness. Normally, under these conditions, the depth modulation of the magnetized zone is determined by the diameter of the recording zone. However, as has been shown for zero bias recording, when the half wavelength of the recorded signal is comparable with the maximum diameter of the recording zone, the magnetization boundary of the zone rotates away from the gap plane (Fig. 3.3). Fig. 3.16a shows the tape magnetization pattern for such a recording with dc bias. The sinusoidal depth modulation of the recording zones is denoted by the dotted line and the magnetization boundary by the solid line. It can be seen that the magnetization boundary lies in the gap plane when the signal change is zero (point A). On the other hand, the magnetization

boundary leads or lags the gap plane when the signal is respectively
rising or falling (points **B** and **C**). The distortion of the recorded
magnetization is therefore a function of the rate of change of magneti-

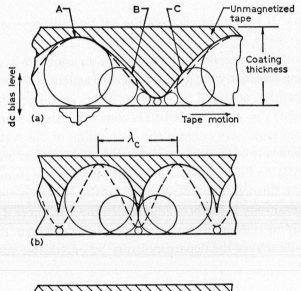

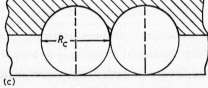

Fig. 3.16.   Tape magnetization patterns, dc bias recording.
Dotted line — locus of magnetization zone diameter.
Solid line   — locus of magnetization boundary.

zation penetration or to the ratio of signal amplitude to wavelength.
For shorter wavelengths (Fig. 3.16b) overlap of the magnetization
boundaries occurs and the minima of the sinusoidal signal are not
recorded at all. In the limit, the magnetization boundary is determined
entirely by the magnetization zone corresponding to the signal maxima.
At this stage (Fig. 3.16c) the character of the recording signal is

entirely lost and is the same for sinusoidal recording or pulse recording with a repetition rate equal to the sinusoidal frequency. This condition may be defined as the resolution limit for dc bias recording (DUINKER [1957]). The magnetization zones will overlap if the tape speed is reduced, or the pulse height increased, or the recording head field function spread longitudinally. Any such change would therefore lead to a reduction in the magnetization changes in the recorded tape. If the critical wavelength for loss-less dc bias recording is λ_c, then

$$\lambda_c = R_c \, , \tag{3.14}$$

where R_c is the diameter of the magnetization zone, given by

$$R_c = 4n(i_b + i_s)/H_c \, . \tag{Eq. 2.4}$$

In practice, using a reproducing head gap length g', the resolution of the reproducing head is determined by the wavelength (λ'_c) at which complete interference takes place in the reproducing head gap, that is, when

$$\lambda'_c = g' \, . \tag{3.15}$$

Thus, the recording resolution is greater than the reproducing resolution if $\lambda < \lambda'_c$ or $R_c < g'$. Using the simple model of cylindrical recording zones,

$$R_{c(max)} = c \, , \tag{3.16}$$

and the recording resolution for maximum signal exceeds the reproducing resolution if the coating thickness, c, is less than the reproducing head gap length. For a finite recording/reproducing head gap length, and considering practical tape hysteresis loops, the recording resolution is a function of the gap length and exceeds the reproducing resolution of the same gap if the ratio of signal to bias is less than 0.1 (DUINKER [1957]).

§ 2.4.   *Pulse Bias Recording*

A modification to dc bias recording having the technical advantage of reducing the bias power is obtained by the use of interrupted dc bias or pulse bias. The duration of the constant amplitude pulses comprising the bias signal may be very short and is limited primarily by the pulse response of the recording head. The repetition rate of the pulses should be high enough to avoid appearance of the pulses in the reproducing signal. Referring to Fig. 3.16, the individual cylindrical magnetization zones would define the magnetized area for pulse biassing. For short wavelength signals, the overlap of magnetization zones tends to mask their discrete occurrences. However, for long wavelength or zero signal, it is necessary for the pulse repetition length to be less than half the resolution of the system to avoid its appearance in the reproducing signal.

§ 3. MISCELLANEOUS MAGNETIC RECORDING METHODS

§ 3.1. *Different Recording Field Directions*

The so-called longitudinal method of magnetic recording has been analysed in detail in the first three sections of this chapter. This method is by far the most successful since it affords the largest wavelength range and the highest resolution with a simple type of recording transducer. Other methods of continuous magnetic storage of information are possible, however, and some have advantages for specific applications, some promise advantages for future recording techniques, while others have been superseded. Various recording methods will be described briefly in this section and their merits assessed.

The simplest transformation from the conventional longitudinal recording head is to change the direction of transport of the magnetic layer to be parallel to the recording gap plane rather than perpendicular to it. In this event, the recorded track width is approximately equal to the recording gap length g. Obviously, however, the resolution is very poor in the direction of motion of the tape, and the method has little application. This recording method is referred to as transverse recording.

Perpendicular recording may be achieved by placing the recording pole pieces on opposite sides of the tape as shown in Fig. 3.17a. Assuming that recording is due to the perpendicular field component only, then recording zones will correspond to the perpendicular equal field contours as shown in the diagram. True perpendicular recording occurs if the tape is transported along the centre line between the poles as indicated by position A. Poor recording resolution results from the large spread of the magnetization zones in this region. Perpendicular

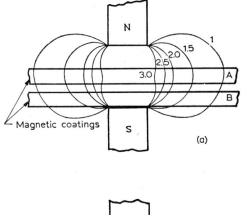

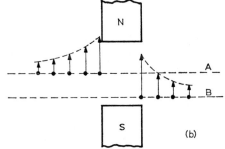

Fig. 3.17.   Perpendicular recording fields:
a. Perpendicular field recording zones.
b. Perpendicular field decrement for tapes (A) and (B).

recording is sometimes used for out of contact recording onto magnetic drums. In this application, one pole only may be used and the magnetic coating may be backed by a high permeability magnetic material which acts as a recording flux sink, resulting in essentially the same field contour as with two poles. Although, at first sight, perpendicular recording appears to have an advantage due to uniform magnetization through the tape coating thickness, similar conditions may be obtained for low resolution recording with a longitudinal recording ring head by using a large gap length in out-of-contact operation.

Better resolution for perpendicular recording results from transporting the tape in contact with one of the pole pieces as shown in position B (Fig. 3.17a). The decrement of the perpendicular field for the two

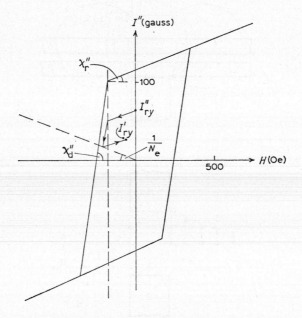

Fig. 3.18.   Demagnetization of perpendicular recording on γ Fe₂O₃ tape.

tape positions is shown in Fig. 3.17b. Although some improvement is obtained for position B, the field falls off slowly compared to the maximum decrement obtainable with longitudinal recording. Using the condition previously assumed that 80% of the irreversible magnetization reversal takes place for reverse fields between $0.2H_s$ and $0.4H_s$, then, if the perpendicular field is made equal to $0.4H_s$ at the trailing edge of the head, irreversible magnetization takes place over a distance equal to $0.25g$ or, say, 0.5 mil. For longitudinal recording, on the other hand, irreversible magnetization occurs in about 0.25 mil at a similar separation from the head. The recording field direction for perpendicular recording turns towards 45° at the trailing edge of the recording pole for tape layers in contact with the pole. Thus, a tape element near the head surfaces experiences a field initially at 45°, rotating towards the longitudinal direction as it moves away from the head. Such rotation of the recording field will cause a loss of the desired perpendicular recording.

Apart from relatively low recording resolution, perpendicular recording has the further restriction of limited long wavelength output. This is caused by the large demagnetizing field effective when the recording medium is magnetized perpendicular to its plane.

Assuming that the practical hysteresis loop model for tape used to describe longitudinal demagnetization (Fig. 2.27) is applicable in perpendicular recording, a remanent magnetization, I''_{ry}, will be reduced to I'_{ry} by the demagnetizing field, as shown in Fig. 3.18. It is seen that the loss depends on the reversible and differential susceptibilities (χ''_r and χ''_d), the slope of the demagnetization line ($1/N_e$), and the reduction of the demagnetizing field due to the reproducing head proximity.

The demagnetizing coefficient will approach $N_e = 4\pi$ at the centre of a tape uniformly magnetized perpendicular to its surface, but will be smaller at the tape edges. The maximum demagnetizing field is shown on the diagram indicating that, for γFe_2O_3 tapes, a maximum remanent magnetization of about 25 % of the retentivity can be obtained for perpendicular magnetization. Since, in this case, a substantial irreversible demagnetization occurs, a reduction in the demagnetization loss will occur if the coercive force is increased. When small demagnetizing coefficients occur, as in longitudinal recording of long wavelengths, the demagnetization line intersects the reversible section of the hysteresis loop, and then the loss depends only on the reversible susceptibility (or permeability). These factors will be taken into account when considering the desirable magnetic properties of tape media (Ch. 5, § 1). The large demagnetizing field in perpendicular recording will also reduce the recording sensitivity considerably. As described in Chapter 2, § 4, the magnetization curve is sheared to a slope of $1/N_e$

both for normal reversible magnetization and for anhysteretic magnetization, as is the case with ac bias recording.

§ 3.2. *Thermoremanent Magnetization*

The recording techniques described in this chapter may all be characterized as coherent magnetization by the signal field with or without the aid of an auxiliary bias field. By far the most successful method for analogue recording is that using ac bias where a highly non-coherent magnetization condition is obtained in the zero signal state. The degree of magnetization randomness achieved for zero signal determines the level of the background noise. In fact, elimination of any dc component in the ac bias is an important factor in obtaining a low noise level.

Other recording techniques produce uniform coherent magnetization for zero signal. Ideally the magnetic potential outside the tape is constant and an external field is obtained when a potential gradient is created by the volume modulation of the tape magnetization. Dc bias recording achieves depth modulation through the tape coating as described in § 2.1. Modulation across the tape width is also possible using a special type of recording head (DANIELS [1952]). Yet another technique in this category uses perpendicular saturation magnetization for bias and longitudinal signal magnetization, resulting in rotation of the saturation magnetization in accordance with the signal (RABE [1960]). Although these last two recording methods have the apparent advantage of good amplitude linearity up to tape saturation, this is offset by the high background noise level obtained due to imperfections in the tape. When the tape is saturated by the bias signal, any magnetic inhomogeneity, either inside or on the surface of the tape, will give rise to local free pole concentrations.

Since a desirable feature of new magnetic recording techniques is the ability to produce all states of magnetization from saturation to complete demagnetization, a promising approach lies in the use of heat energy to produce the demagnetized condition. If the recording medium is heated above its Curie point, ferromagnetism is completely destroyed. If, in addition, cooling takes place in the absence of any external magnetic field, a highly random magnetization state is acquired, determined only by the randomness of the internal fields. The acquisition of coherent magnetization on cooling from above the Curie point, in the presence of a small applied field, will now be considered with a view to its application to a magnetic recording process. The important factors to be determined are the magnetization susceptibility, linearity,

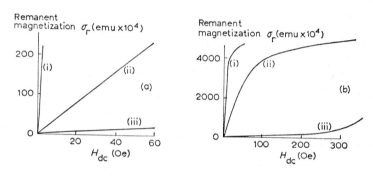

Fig. 3.19.   Remanent magnetization curves (Rimbert [1957]):
a.   Dilute γ Fe$_2$O$_3$;   b. Dilute Fe$_3$O$_4$.
(i)   Thermoremanent magnetization.
(ii) Anhysteretic remanent magnetization.
(iii) Isothermal remanent magnetization.

and stability. The process to be considered is similar to the acquisition of magnetization by rocks in the earth's field; indeed, some of the constituents of rocks are similar to the ferrimagnetic oxide powder used in recording tapes. The analysis of the thermal magnetization process in single and multidomain particles has been reviewed by Néel [1955].

It has been shown experimentally (Rimbert [1957]), that the magnetization acquired by heating a magnetic material above its Curie point in the presence of a field (H) which is small compared to the room temperature coercivity of the material, is large compared to that acquired normally or anhysteretically at room temperature. Fig. 3.19a shows the three magnetization curves for small dc fields (H_{dc}) using γ Fe$_2$O$_3$, diluted with a nonmagnetic powder. The corresponding curves for Fe$_3$O$_4$ are shown for dc fields sufficient to achieve anhysteretic remanence. These results indicate that the use of heat rather than an ac field results in an increase in magnetization susceptibility whilst apparently retaining a similarly linear characteristic. Also, as with anhysteretic magnetization, the magnetization achieved by simultaneous application of heat is much less susceptible to destruction by subsequently applied magnetic fields or heat.

For particles exhibiting shape or crystal anisotropy, the coercivity is temperature dependent and becomes zero at the Curie point. In addition, the internal fields, being proportional to the intrinsic magnetization, also approach zero near the Curie point. Thus the ability of an applied small field irreversibly to magnetize particles increases with temperature. Near the Curie point a very small field can magnetize all the particles. As the temperature falls, the intensity of magnetization of the particles increases, and at room temperature the maximum possible value is approached.

This simple picture of infinite thermomagnetization susceptibility is, however, not borne out in practice although, as shown in Fig. 3.19, very high susceptibilities are obtained by this method. In order to account for the finite susceptibility, it is necessary to consider the manner in which magnetization is acquired during the temperature decrement. This in turn will depend on the relative magnitudes of internal fields, thermal disturbances, and switching fields at elevated temperatures.

As the temperature is reduced below the Curie point, the magnetization acquired tends to be made random by the thermal agitation (NÉEL [1955]). The thermal energy required to perturb the magnetization direction irreversibly is proportional to the particle switching field, spontaneous magnetization, and volume. Thus the large particles will be the first to be frozen in their direction of magnetization as the temperature is reduced. Magnetization thus acquired at high temperatures is very stable at room temperature, and only those particles of small volume which become frozen near room temperature are unstable. Such small particles are usually avoided in magnetic tape powders since their susceptibility to irreversible magnetization by small fields near room temperature gives rise to unwanted recording, called "print through", from adjacent recorded tape layers (see Ch. 4 § 7). The principle of heating a material above its Curie Point to achieve magnetization change has been used in "Curie Point Writing" (MAYER [1958]).

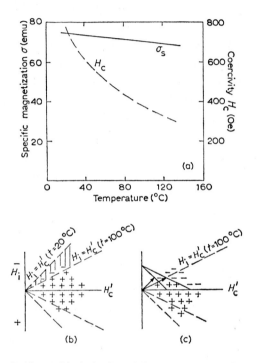

Fig. 3.20.   Magnetization of cobalt doped iron oxide at elevated temperatures:
　　a. Temperature dependence of H_c and σ_s for cobalt doped iron oxide.
　　(see Fig. 6.8)
　　b. Preisach diagram to illustrate magnetization change on heating.
　　c. Magnetization due to small field at 100°C.

For continuous magnetic recording, the high temperatures required to achieve Curie Point Writing may be inconvenient for many applications. However, in some ferrites it is not necessary to reach the Curie point to achieve irreversible magnetization changes. The extent of the change of magnetization depends on the modes of variation of the spontaneous magnetization and the switching fields with temperature. This is illustrated by the experimental curves of the temperature dependence of coercivity and magnetization shown in Fig. 3.20a, for a cubic ferrite whose magnetic properties are controlled by crystal anisotropy. It is seen that a minor decrease of specific magnetization σ_s occurs on heating to 100° C, whereas H_c falls to about half of its room temperature value. Thus, on heating, the internal fields remain unchanged while the particle switching fields reduce and demagnetization occurs, which remains after cooling. The irreversible loss of magnetization on heating such materials may be explained with the aid of a Preisach diagram (Fig. 3.20b). Reduction of switching fields accompanied by non-variant internal fields may be simulated by compression of the particle distribution function along the abscissa towards the origin, or by reducing the slope of the lines $\pm H_i = H'_c$. The particles contributing to the overall demagnetization are shown in the shaded area. The rate of change of magnetization with temperature will therefore depend on the distribution $I(H_i, H'_c)$ as well as on the temperature dependence of (H'_c). Such materials could be used in several ways for recording. Obviously, recording by heat erasure of a presaturated tape is possible. The mechanism would differ from the Curie Point Writing technique since erasure is due to internal rather than external demagnetizing fields. These materials will also have increased susceptibility to magnetization by small fields at elevated temperatures as shown by Fig. 3.20c, although the amplitude linearity would be poor at low levels.

Examples of possible techniques for magnetic recording with the aid of heat have been discussed to illustrate their promising aspects. Future developments along these lines may be expected to lead to new recording techniques which could compete with existing methods.

CHAPTER 4

MAGNETIC REPRODUCING PROCESS

§ 1. INTRODUCTION

The general objective of the reproducing process is to detect faithfully the recorded magnetization in the tape, without producing any permanent change in the tape magnetization. Faithful detection implies the production of an electrical signal proportional to the recorded magnetization pattern for the whole range of recorded wavelengths. Herein lies a fundamental limitation of the longitudinal recording process at long wavelengths; as the wavelength becomes very long, the surface field approaches zero and the recorded magnetization is normally undetectable by any transducer which measures the field external to the tape. Practical tapes are not perfectly homogeneous, however, and a small external field does exist which may be detected as noise: this noise then increases as the tape magnetization increases. Where it is necessary to record and reproduce very low frequencies it is more efficient to use perpendicular or transverse recording techniques where the magnetic poles of the recording are distributed on the tape surfaces or edges for infinitely long wavelengths, thus producing a detectable field. Amplitude linearity is a further necessary feature for reproduction of analogue type recordings. Whereas this limits the field detection devices suitable for such recordings, other limitations, such as available signal-to-noise ratio and resolution, often restrict the use of many transducers even in binary type recording where linearity is not important.

The reproduction transducers used almost universally in magnetic recording systems rely on the magnetization of high permeability polished pole-pieces, defining a short air gap parallel to the trailing edge of the recording head, and held in intimate contact with the

85

moving tape coating. A low reluctance core shunts the pole-piece magnetization in the well-known ring head illustrated in Fig. 4.1. Thus the recorded flux is effectively directed through the core (path c) as shown, although some will pass through undesirable paths between

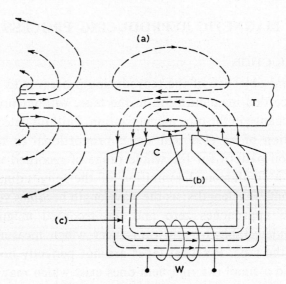

Fig. 4.1.   General disposition of reproducing head core C, winding W, and
recorded tape T.
External closure paths for recorded flux lines:
(a) Above coating.
(b) Between head and coating.
(c) Through head core.

the head and the coating (path b) and on the remote side of the tape (path a).

Detection of the core flux is a fairly straightforward matter and will not be dealt with at length in this chapter. It is assumed that the conventional method of detecting the core flux by a multiturn winding is used as shown in Fig. 4.1. In this case, the voltage across the winding is proportional to the time derivative of the core flux; thus the head will deliver zero output for zero frequency as well as for infinite wavelength as described earlier. However, if desired, various techniques are available to detect a nonvarying core flux. This may be achieved directly by opening the core circuit and inserting a small field detector utilizing the phenomena of magnetoresistance, magnetic resonance, or Hall effect, for instance. Another successful method involves periodic interruption of the core flux, thus allowing its detection by the conventional winding. This could be achieved mechanically, but is obtained more elegantly by periodic saturation of the core, producing variable reluctance. Saturation is achieved in a symmetrical fashion, for instance by a transverse field, so as to avoid its direct detection by the pickup winding and also to prevent any flux appearing across the pole-piece gap.

The practical configuration of the pole-piece surfaces which contact a moving tape is shown in Fig. 4.2a. Two sets of poles are shown, having collinear gaps. Magnetic shields between the pole-piece sets help to minimize the spread of tape flux from one set of poles to another. As indicated in Fig. 4.1, the flux closure paths external to the tape are modified when the tape is in the vicinity of the reproducing pole-pieces. This action is clearly illustrated in Fig. 4.2b where a prerecorded tape is placed in contact with the reproducing heads of Fig. 4.2a; the surface fields in path 'a' above the tape are made visible with a magnetic powder indicator. The short wavelength resolution of the reproducing head is determined by the air gap length between the pole-pieces, and the long wavelength limit by the length of the pole-pieces.

Due to the high degree of mechanical perfection achievable in polished flat pole-piece edges, very small gaps can be produced yielding the highest reproduction resolution. Other methods of direct detection of the surface field of the recording, including deflection of an electron beam directed near the recorded coating surface and magneto-optical rotation of the plane of polarization of a light beam reflected off the tape, have found only limited application, see e.g. FREUNDLICH et al. [1961], LENTZ and MIYATA [1961].

The scope of this chapter covers the analysis of the process of tape magnetization detection by high permeability pole-pieces and the efficiency of the core circuit in capturing the available pole-piece flux. The following four sections are devoted to this process for recorded signal detection. In the final section the reproduction of tape noise is considered for the same transducer system.

§ 2. REPRODUCTION OF LONGITUDINAL RECORDING WITH GAPPED RING HEAD

§ 2.1. *Relation Between Tape Flux and Reproducing Head Flux*

The reproducing head to be considered in the analysis of the reproducing process is similar to that described in the introduction, with the added restrictions that the permeability of the core is considered to be infinitely high, the depth of the gap to be very large and the poles to have infinite length in contact with the tape. The modifications necessary for consideration of finite permeability are described later, and shown to be relatively small. The general technique used to determine the flux produced in the reproducing head by the recorded tape magnetization is to apply the appropriate sensitivity factor to each element of tape in the vicinity of the gap. Since the reproducing process is essentially a linear one, it is valid to calculate this reproduction sensitivity function by determining the field around the gap when the head is energized. In other words, a reciprocal relationship occurs and the field distribution in the gap region may be looked upon as a measure of the relative spatial sensitivity of the head to magnetization in the gap region.

The longitudinal field at the point x, y, z, due to a current, i_c, in the head winding would produce a voltage, Δe_a, in a single turn wrapped around the elemental volume dx, dy, dz, as shown in Fig. 4.3. Thus

$$\Delta e_a = - \, d(H_x \, dy)/dt \qquad (4.1)$$

considering unit elemental tape width. It is assumed that, since the tape width, w, is normally large compared to the gap length g' and the coating thickness, c, the z variations of tape magnetization and head sensitivity are negligible and the problem is accurately described by a two-dimensional calculation.

Shields

Pole pieces

Fig. 4.2a.   Multitrack reproducing head pole-pieces with collinear gaps.

Fig. 4.2b. Recorded tape placed on multitrack head. External field (path a, Fig. 4.1.) made visible using magnetic powder indicator.

In addition, the longitudinal tape magnetization, I_x'' for the element dx, dy, dz, is a magnetic shell which can be expressed in terms of an equivalent current, i_a, in a single turn encompassing the element, where i_a is the strength of the shell. Hence,

$$i_a = \frac{\text{magnetic moment of element}}{\text{area}}$$

$$= I_x'' \, dx \, . \tag{4.2}$$

Now the current, i_a, produces a voltage, $\varDelta e_c$ in the head winding. Hence, by reciprocity

$$\varDelta e_c / \varDelta e_a = i_a / i_c \, . \tag{4.3}$$

Combining equations (4.1), (4.2) and (4.3) yields a relationship for the head core flux, $\varDelta \varphi_c$, due to the element of magnetization $I_x'' \, (x, y)$ in terms of the relative field at x, y. Integrating over all contributing elements gives

$$\varphi_c = \frac{2\pi}{U_0} \int\limits_{a}^{a+c} dy \int\limits_{-\infty}^{+\infty} H_x I_x'' \, dx, \tag{4.4}$$

where φ_c is the head flux per unit tape width, and $U_0 = 2\pi i_c =$ potential of pole pieces due to current, i_c, in the head winding.

In order to determine the head core flux from eq. (4.4), it is necessary first to establish the field function in the vicinity of the gap, $H_x(x, y)$, on passing a current through the head winding. This is considered next, and is followed by a calculation of the reproducing head flux corresponding to a sinusoidaily magnetized tape.

§ 2.2.   *Reproducing Head Field Function*

In this section, the magnetic potential distribution $U(x, y)$ around the gap will be determined, the region of interest being that occupied by the tape. In order to solve this problem it will be necessary to consider both the potential distribution in the x, y plane above the gap [i.e. $U_a(x, y)$; y positive and $-\infty < x < +\infty$] and also the potential in the gap [$U_g(x, y)$; y negative $-1 < x < +1$]. It is assumed that the gap width (z dimension) is large compared to its length and that a two dimensional plot in the x, y plane, with the pole faces considered as equipotential surfaces, represents closely the conditions occurring in practice. Once the potential distribution $U_a(x, y)$ has been determined the corresponding field distribution may be obtained by differentiation. Thus,

$$H_x = - \partial U_a / \partial x . \tag{4.5}$$

Several methods exist for the determination of the potential distri-

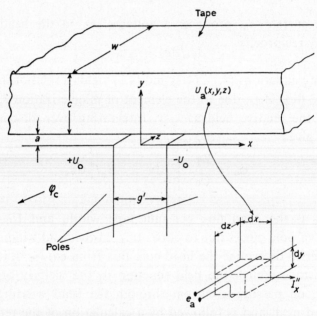

Fig. 4.3.   Reproduction with gapped ring head.

bution around the gap. For instance, the method of conformal mapping has been used to determine this distribution by graphical methods (BOOTH [1952], WESTMIJZE [1953a], KARLQVIST [1954], GREINER [1953, 1955, 1956]). This technique is applicable only in the case of unit permeability media outside the gap. Another method, using Fourier transformations, allows a continuous analytic expression for the field distribution in media of any permeability to be obtained (KARLQVIST [1954], SCHWANTKE [1957], FAN [1961a]): the last derivation is considered here.

It is necessary that the solution obtained for the potential obeys Laplace's equation for the two dimensional case,

$$\frac{\partial^2 U}{\partial x^2} + \frac{\partial^2 U}{\partial y^2} = 0 , \qquad (4.6)$$

and also that it satisfies the boundary conditions that the pole faces are equipotential surfaces $\pm U_0$. Since, for the gap shown in Fig. 4.3, the boundaries correspond to coordinate lines in the x, y plane, a solution of eq. (4.6) by separation of variables is the most convenient method. Thus,

$$U(x, y) = X(x) \, Y(y) , \qquad (4.7)$$

where X is a function of x alone and Y is a function of y. Insertion in eq. (4.6) leads to the ordinary differential equations,

$$\left.\begin{aligned} \frac{d^2 Y}{dy^2} - k^2 Y = 0 \\[2em] \frac{d^2 X}{dx^2} + k^2 X = 0 \end{aligned}\right\} , \qquad (4.8)$$

having the general solutions

$$Y = A_1 \exp(ky) + B_1 \exp(-ky) , \qquad (4.9a)$$

or equivalently

$$Y = A_2 \sinh ky + B_2 \cosh ky ,$$

where k is a constant,
and

$$X = C_1 \sin kx + D_1 \cos kx , \qquad (4.9b)$$

where A_1, A_2, B_1, B_2, C_1, and D_1, are constants to be determined by the boundary conditions. The product solution obtained from eqs. (4.7), (4.9a) and (4.9b) may be written in terms of an integral for the semi-infinite region above the pole faces (y positive), and as a summation in the gap region. In this way the boundary conditions are satisfied in these two regions with allowed values of K being continuous and discrete, respectively. Thus, due to symmetry, attention may be limited to positive values of x, giving

$$U_a(x, y) = \int_0^\infty A(k) \sin kx \exp(-ky)\, dk,$$ (4.10)

$$(y > 0, \quad 0 < x < \infty)$$

$$U_g(x, y) = U_0 x + \sum_{n=1}^\infty C_n \sin(n\pi x) \exp(n\pi y).$$ (4.11)

$$(y < 0, \quad 0 < x < 1)$$

These general solutions have been worked out in the following manner by FAN [1961a]. Eqs. (4.10) and (4.11) may be combined by assuring continuity of the potential at $y = 0$. Also, if the tape permeability is unity the y derivative of the potential at $y = 0$ is continuous, leading respectively to

$$\left.\begin{aligned} \int_0^\infty A(k) \sin kx\, dk &= U_0 x + \sum_{n=1}^\infty C_n \sin(n\pi x) \\ &\quad (0 < x < 1) \\ &= U_0 \ (1 < x < \infty) \end{aligned}\right\},$$ (4.12)

and

$$\int_0^\infty A(k) k \sin kx\, dk = - \sum_{n=1}^\infty C_n n\pi \sin(n\pi x).$$ (4.13)

The right hand side of eq. (4.12) indicates that, along the line $y = 0$, the potential drop from the left to right hand pole-piece is the summation of a linear x dependent term and a Fourier sine series. It will be shown later that the first three harmonic components of the series are sufficient to describe accurately the practical conditions. However, earlier approximations, using the first term only, do lead to significant errors.

Performing a Fourier sine transformation on eq. (4.12) and operating on eq. (4.13) to determine the Fourier coefficients allows the elimination of the coefficient $A(k)$ and the determination of coefficients C_n. The latter converge rapidly, and the first three are:

$$C_1 = -0.082\, U_0; \quad C_2 = 0.027\, U_0; \quad C_3 = -0.014\, U_0. \quad (4.14)$$

Finally, the required potential distribution $U_a(x, y)$ in the region of the tape may be written in terms of the known coefficients C_n by substituting for $A(k)$ in eq. (4.10) from the transformation of eq. 4.12 giving,

$$U_a(x, y) = \frac{2U_0}{\pi} \int_0^\infty \frac{\sin k \sin kx}{k^2} \exp(-ky)\, dk$$

$$+ \sum_{n=1}^\infty C_n\, 2n(-1)^n \int_0^\infty \frac{\sin k \sin kx}{k^2 - (n\pi)^2} \exp(-ky)\, dk, \quad (4.15)$$

from which the x and y components of the field may be obtained by differentiation (eq. 4.5).

§ 2.3.   Reproducing Head Core Flux

From the reciprocity calculation, the flux in the reproducing head core is given as a function of the field distribution around the gap when the head is excited by a current in its winding. If the recorded tape is magnetized sinusoidally in the x direction only, it is the x component of the gap field distribution which acts as the weighting factor in the reproduction process. Thus from eqs. (4.5) and (4.15)

$$H_x = \frac{2U_0}{\pi} \int_0^\infty \frac{\sin k \cos kx}{k} \exp(-ky)\, dk$$

$$+ \sum_{n=1}^\infty C_n\, 2n(-1)^n \int_0^\infty \frac{k \sin k \cos kx}{k^2 - (n\pi)^2} \exp(-ky)\, dk. \quad (4.16)$$

On transporting the sinuoidally magnetized tape past the head at a velocity, v, the instantaneous magnetization is given by,

$$I_x'' = I_{x(\text{max})}'' \cos\left[\frac{2\pi(x - vt)}{\lambda}\right]. \qquad (4.17)$$

However, by symmetry, the only component of eq. (4.17) to contribute to the reproducing head flux is

$$I_x'' = I_{x(\text{max})}'' \cos\left(\frac{2\pi x}{\lambda}\right)\cos\left(\frac{2\pi vt}{\lambda}\right). \qquad (4.18)$$

Combining eqs. (4.4), (4.16) and (4.18) leads to an expression for the reproducing head flux, φ_c, in terms of the tape magnetization, I_x'', tape thickness, c, head to tape spacing, a, and the recorded wavelength, λ,

$$\varphi_c = 4\pi c\, I_{x(\text{max})}'' \cos\left(\frac{2\pi vt}{\lambda}\right)$$

$$\times \left[\left\{1 - \exp\left(-\frac{2\pi c}{\lambda}\right)\right\}\Big/\left(\frac{2\pi c}{\lambda}\right)\right]$$

$$\times \left[\exp\left(-\frac{2\pi a}{\lambda}\right)\right]$$

$$\times \left[\left\{\sin\frac{2\pi}{\lambda}\Big/\left(\frac{2\pi}{\lambda}\right)\right\}\left\{1 + \sum_{n=1}^{\infty} \frac{A_n}{U_0}\frac{4\pi n(-1)^n}{4 - n^2\lambda^2}\right\}\right] \qquad (4.19)$$

whence the reproducing head voltage per turn may be obtained by differentiation.

§ 2.4.   *Reproduction Loss Factors*

Each of the square bracketed factors in eq. (4.19) has a wavelength dependent effect on the response of the reproducing head. Hence the reproducing head flux for a tape magnetized sinusoidally along its length and uniformly through its thickness will be correspondingly sinusoidal but with its amplitude modified by the following factors, (a), (b) and (c).

$$\text{(a)} \quad \left[\left\{1 - \exp\left(-\frac{2\pi c}{\lambda}\right)\right\}\bigg/\left(\frac{2\pi c}{\lambda}\right)\right].$$

This factor, determined only by the ratio of the coating thickness to the

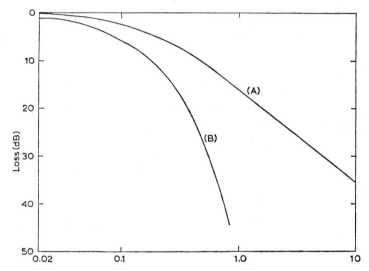

Fig. 4.4.   Monotonic wavelength dependent losses in reproduction. Curve A (coating thickness/wavelength c/λ) denotes thickness loss. Curve B (head to tape spacing/wavelength a/λ) denotes spacing loss.

recorded wavelength, approaches a maximum value of unity when the wavelength is long compared to the coating thickness. However, when the wavelength is short compared to the coating thickness, the factor tends to zero producing maximum attenuation of the reproducing head flux. The term may be viewed as a tape coating thickness loss, L_c, given by

$$L_c = 20 \log_{10}\left[\frac{2\pi c/\lambda}{1 - \exp(-2\pi c/\lambda)}\right]. \qquad (4.20)$$

This loss is plotted in Fig. 4.4 curve A, and refers to the attenuation

of the reproducing head flux relative to the maximum flux obtained
when the wavelengths are long compared to the coating thickness. It
can be seen that for a wavelength range of 1:200, normally obtained
in audio recording, the reproducing head flux is attenuated by a factor
of at least 30 times if the coating thickness is sufficient to produce the
maximum flux at the longest wavelength.

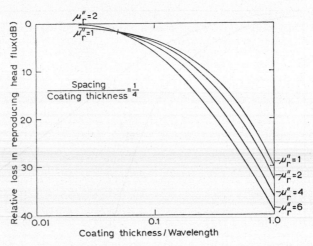

Fig. 4.5. Reproducing head flux as a function of recorded wavelength (parameter, reversible permeability). Relative spacing constant. (WESTMIJZE [1953b]).

The thickness loss can be accounted for by considering the action of
the demagnetizing field of the recording during the reproduction
process. As shown in Fig. 4.1. the demagnetizing field external to the
tape is equal on both sides of the coating when the tape is not in
contact with the head, assuming uniform magnetization through the
coating thickness. Inside the tape the demagnetizing field produces a
reversible and irreversible loss of magnetization: the former is of some
significance in the reproducing process since the partial recovery of the
reversible loss on contacting the reproducing head is a factor in the
total thickness loss. When the recording is in the vicinity of the reproducing head, the demagnetizing field is redistributed and is mainly
active in magnetizing the reproducing head core. As can be seen in

Fig. 4.1, the external field on the remote side of the tape (path a) is reduced while that on the near side is increased. Also, the internal field is reduced, particularly on the near side. For sufficiently long wavelengths and thin coatings the external tape field will exist entirely on the near side and no thickness loss will occur. This condition is approached in Fig. 4.2, showing little external field from the recording in the vicinity of the pole tips. For shorter wavelengths however, an increasing proportion of the demagnetization field will act in path a, and in the tape layers remote from the head. Thus, a loss of near side field will occur with consequent loss of reproducing head flux. In addition, the increasing demagnetizing field with distance from the head can cause a reduction of magnetization in the remote layers if the reversible permeability, μ_r'', of the tape is greater than unity. This will lead to further reproducing head core flux loss below the maximum attainable value. In keeping with the above description, Fig. 4.5 shows the dependence of reproducing head core flux on relative recorded wavelength for tapes of different reversible permeabilities (WESTMIJZE [1953b]). It is assumed that an infinitely short gap length exists in the reproducing head for the calculation leading to Fig. 4.5. The effect of tape permeability on the calculated reproduction function (eq. 4.19) is considered in the next section.

$$(b) \quad [\exp(-2\pi a/\lambda)] \, .$$

This factor, determined by the ratio of the tape spacing to the recorded wavelength, may be expressed as a spacing or separation loss, L_a, by

$$L_a = 20 \log_{10} \exp(2\pi a/\lambda)$$

$$= 54.6a/\lambda \, . \tag{4.21}$$

which is plotted in Fig. 4.4 – curve **B**.

In Fig. 4.1 the external flux closure path in between the tape and the reproducing head (path b), accounts for the spacing loss. For tapes with reversible permeability greater than unity, the head to tape spacing will also influence the magnetization as a function of depth into the coating since the internal demagnetizing field will increase with spacing, especially in the near surface layers. Hence, short wavelength magnetization will be the more attenuated the greater the reversible permeability and the greater the spacing. Fig. 4.6 shows the dependence of the reproducing head flux on the recorded wavelength for different spacings, illustrating the interdependence of thickness and spacing loss described (WESTMIJZE [1953b]).

When very short wavelengths are recorded, it is the spacing loss which provides the major attenuation of the reproducing head flux and

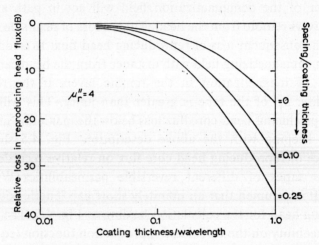

Fig. 4.6.   Reproducing head flux as a function of recorded wavelength (parameter, relative spacing). Reversible permeability constant. (WESTMIJZE [1953b]).

is, then, a determining factor in the resolution of the system. For this reason tapes are often polished or calendered to minimize the spacing loss. The severity of this loss is indicated in the examples of high resolution reproducing conditions listed in Table 4.1.

TABLE 4.1.

Separation losses in reproduction

System	Frequency	Tape speed (inch/sec)	Separation loss (dB per 0.1 mil)
Professional audio recording	15kc/s	15	5.5
High resolution audio recording	15kc/s	$1\frac{7}{8}$	44.0
Direct video recording	3Mc/s	200	81.5

In spite of polished tape surfaces to minimize spacing loss, there will be some inherent separation due to the reduction of the reproducing head permeability in the surface layers of the pole-pieces and due to the particulate nature of the magnetic component of the coating. The latter factor, determined by the degree of dispersion obtained in the coating, can give rise to a noise which is dependent on the degree of

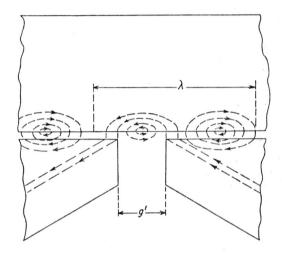

Fig. 4.7.   Reproduction of short wavelength recording, where $\lambda \approx 2g'$.

magnetization of the tape. The role of separation loss in tape noise is discussed further in Section 6 of this chapter.

$$(c) \quad \left[\sin (\pi g'/\lambda)/(\pi g'/\lambda)\right]\left[1 + \sum_{n=1}^{\infty} \frac{A_n}{U_0} \frac{4\pi n(-1)^n}{4 - (2n\lambda/g')^2}\right].$$

This factor, referring to a gap length, g', accounts for the effect of the finite gap length on the reproducing head flux. As shown in Fig. 4.7, when the recorded wavelength is about twice the gap length, some of the flux external to the tape can close in the gap region without passing round the reproducing head core, thus constituting a loss of reproducing head core flux. As the recorded wavelength becomes shorter, destructive interference occurs between oppositely magnetized elements of tape situated in between the pole-pieces, leading to a series of minima in the core flux. The first term, $\sin(\pi g'/\lambda)/(\pi g'/\lambda)$, has been obtained in simpler derivations of the reproducing process e.g. WALLACE [1951]. This accounts for minima in the core flux when

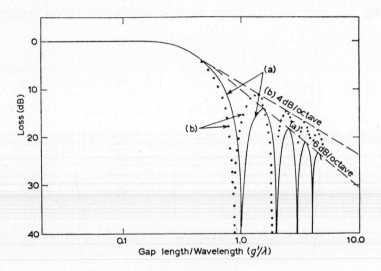

Fig. 4.8.   Gap loss function in reproduction.
(a) Simple formula.
(b) Complete formula.

the gap length is an integral number of wavelengths, as shown in Fig. 4.8 – curve a, where the corresponding approximate gap loss is given by

$$L_g' \approx 20 \log_{10}\left[\frac{\sin(\pi g'/\lambda)}{\pi g'/\lambda}\right]. \qquad (4.22)$$

In order to preserve a continuous wavelength response in reproduction, it is necessary then that the first minimum in the response (maximum in attenuation) occurs outside the wavelength range of interest. This is achieved by ensuring that the gap length is shorter than the shortest recorded wavelength. In the range of $g' < \lambda_{min}$ a first approximation to the second term in (c), containing the series, is given by taking into account $n = 1$ only. In this case

$$L_g' = 20 \log_{10}\left[\left\{\sin(\pi g'/\lambda)/(\pi g'/\lambda)\right\}\left\{\frac{5 - 4(\lambda/g')^2}{4 - 4(\lambda/g')^2}\right\}\right]. \quad (4.23)$$

The essential modification to the simple function, shown in Fig. 4.8 – curve a, is that the first maximum attenuation occurs at a somewhat longer wavelength where $\lambda = g'/0.88$ (FAN [1961a]). This result is in agreement with experiment (DANIEL and AXON [1953]), and with other analyses of the reproducing process which take account of the spatial sensitivity function of the reproducing head (WESTMIJZE [1953a]). In addition, if further terms of the series are taken into account, the gap loss function may be accurately assessed for shorter wavelengths than $\lambda = g'$. In contrast to the simple formula, successive minima in the gap loss (maxima in Fig. 4.8) lie on a line of 4 dB per octave slope as shown in Fig. 4.8 – curve b. This latter result is also in agreement with experimental curves (DANIEL and AXON [1953]), lending strong support to the fundamental accuracy of the reproduction performance predicted by eq. (4.19).

If the reversible permeability, μ_r'' ,of the tape is greater than unity, the calculation of the gap region potential function must be modified to account for the jump in the normal field component by a factor μ_r'' at the near tape surface. This correction enters in eq. (4.13) giving modified values for the constant C_n such that the numerical values of C_1, C_2 etc. increase with μ_r'' and are approximately doubled for $\mu_r'' = \infty$ (FAN [1961a]). Thus the series term in eq. (4.19) is increased, leading to a further shift (about 10%) of the response minima due to gap loss.

§ 2.5.   *Comparison with Practical Conditions*

The model for reproduction of a sinusoidally recorded tape with a ring type head agrees quite well with experimental conditions. The major assumptions are constant magnetization with depth into the coating, unit reversible tape permeability, and infinite permeability in the reproducing head pole-pieces. In addition, only the longitudinal component of the recorded magnetization is considered. These assumptions will be further examined to establish their range of applicability.

The first two assumptions are interrelated since a reversible permeability greater than unity leads to a reversible magnetization loss, due to the demagnetizing field of the recording, which is greater at the centre of the coating than at its surfaces. From Fig. 2.8 (Ch. 2, § 3) it can be deduced that the reversible susceptibility, χ_r'', parallel and perpendicular to the direction of orientation has approximate values 0.08 and 0.12, respectively, and is, to a first approximation, independent of magnetization level. In the reversible magnetization regions of the hysteresis loop the reversible permeability is given by

$$\mu_r'' = 1 + 4\pi\chi_r'' \tag{4.24}$$

giving typical values, for γ Fe_2O_3 powder tape, of $\mu_r'' = 2.0$ and 3.0 parallel and perpendicular to the orientation direction, respectively. Hence, considering the effects of $\mu_r'' = 2.0$ on the thickness, spacing, and gap losses already described, it is seen that the resulting reproducing head flux is little changed from that calculated for unit permeability. Further support for the premise that little reversible magnetization loss occurs at medium and long wavelengths is obtained from the experimental result that reproduction by means of a short gap head, of the type described here, and by means of a non-magnetic wire are virtually identical after corrections have been made for their respective geometry-dependent losses (DANIEL and AXON [1953]). This would not be so if the reduced internal tape demagnetizing field on contacting the pole-pieces produced significant increase of tape magnetization.

At short wavelengths greater demagnetizing fields are active; it is then necessary that the magnetization be kept at a low enough level to avoid irreversible demagnetization losses since these lead to distortion of large amplitude recording at short wavelengths. Even for short wavelengths, self demagnetization losses due to reversible effects are probably small in γ Fe_2O_3 tapes. It is important in the design of tapes having high output at short wavelengths to ensure that the reversible permeability be as small as possible, and also that the coercive force is sufficiently large to avoid excessive irreversible demagnetization loss.

Although it is known that the magnetization achieved in recording is not uniform with depth at short wavelengths, due to the reduction of recording resolution with separation from the recording head, this is possibly not a major factor in determining the reproducing head flux. This viewpoint is conjectured on the wavelength dependent thickness and separation losses in reproduction being more severe than the corresponding reduction of recorded magnetization with depth in the coating. Support for this viewpoint is obtained from analyses of the reproducing head output for constant signal amplitude recordings over a large range of recorded wavelengths. It is found that the thickness, spacing, and gap loss formulae (eq. 4.19) can account for the experimental output vs. wavelength curves when reasonable values are assumed for spacing and gap length (WALLACE [1951], COMERCI [1962]).

The initial permeability of nickel-iron alloys used in reproducing heads may be as high as $\mu_c = 30\,000$ although, in practice, this value could not be expected to exist in the gap region due to mechanical polishing stresses during construction and stresses caused by head-to-tape friction. Also, high frequency applications in the megacycle range entail further reductions of the effective permeability due to losses in the core material. It is, therefore, more realistic to consider that the head initial permeability is of the order of 1 000 for many high frequency applications. Soft ferrites are sometimes used as head cores for high frequency reproduction and these too have initial permeabilities in the region of 1 000. The effect of moderate head permeabilities (> 100) on the reproducing performance has been studied by considering the potential distribution in the head poles in addition to that considered previously in the gap region only (FAN [1960a]). Eddy current effects will tend to concentrate the flux in the surface layers so that the pole-pieces act more like semi-infinite sheets in the x-z plane. However, this type of head has been shown to behave somewhat similarly to the model considered (WESTMIJZE [1953a]). The overall effect of finite permeability pole-pieces appears to be quite small and for most practical cases it is sufficiently accurate to assume $\mu_c = \infty$.

Finally, the analysis of the reproducing process, outlined in this section, has been confined to longitudinally magnetized tape. It was shown in Chapter 2, however, that the perpendicular component of the recording field is large near to the pole-pieces. Although the dominant effect on the recorded magnetization is one of attenuation, the perpendicular component of magnetization in longitudinally oriented tapes has been estimated to have a maximum value about 10% of the total magnetization. In the reproducing process this component can be accomodated quite easily into the previous analysis by reciprocal application of the perpendicular field function H_y. If this is combined with the perpendicular magnetization, I_y'', in a reciprocity equation analogous to eq. (4.4) for I_x'', the wavelength dependence of the reproducing head flux on the perpendicular magnetization component is obtained. It is found that this is identical to the longitudinal component wavelength dependence apart from a phase shift of $90°$. Hence the perpendicular component of the tape magnetization may be accounted for in eq. (4.19) by a change of phase and maximum amplitude of the reproducing head flux. For unit tape permeability a rotation of the recorded magnetization from longitudinal to perpendicular, at constant amplitude, produces only a change in phase in the reproducing head core flux. However, if the permeability is greater than unity, the perpendicular component suffers more demagnetization loss than the longitudinal component for long wavelengths. When the wavelength is short compared to the coating thickness, the tape magnetization does not depend on its direction.

§ 3. REPRODUCTION OF PERPENDICULAR RECORDING WITH SINGLE CONTACTING POLE

It was shown in Chapter 2 that 'perpendicular' recording achieved by transporting a tape in contact with a single recording pole-piece is essentially a low resolution process involving small magnetization levels. Also, due to the rotation of the recording field towards the longitudinal direction, the recording is not entirely perpendicular, particularly in the surface layer of the tape in contact with the pole. Reproduction of true perpendicular recording with a similar single pole in contact with the tape is now considered. The pole and tape dimensions are illustrated in Fig. 4.9. The reproducing head core is not shown, but can be imagined to link the contacting pole and a non-contacting pole on the remote side of the tape in the plane $x = 0$ at unit distance from the contacting pole.

The analysis of the reproducing process is entirely analogous to that described in the previous section. The appropriate solution to Laplace's equation is obtained by applying the boundary conditions to eq. (4.9). In this case, it is noted that the pole geometry is similar to that considered for the positive values of x for the pole-pieces of Fig. 4.3, except that the coordinates x and y are interchanged, and the analogue of the gap depth is $2L$, rather than infinity, as shown in Fig. 4.9. The analogous equations to (4.10) and (4.11) for zones A and B are (FAN [1960b])

$$U_B(x,y) = \int_0^\infty A'(k) \sin ky \exp(-kx) \, dk \, ,$$
$$(x > 0, \quad 0 < y < \infty) \tag{4.25}$$

$$U_A(x,y) = U_0 y + \sum_{n=1}^\infty C_n' \sin(n\pi y) \cosh \lambda \, n\pi (L + x) \, ,$$
$$(0 > x > -L, \quad 0 < y < 1) \, . \tag{4.26}$$

Operation on eqs. (4.25) and (4.26) in a similar manner to that described in the previous section leads to the potential distribution around the reproducing head in terms of known coefficients C_n'; the field distribution may be obtained by appropriate differentiation. The tape magnetization is again assumed to be uniform through the thickness of the tape. Hence, the reproducing head core flux for perpendicularly magnetized tape may be calculated in terms of the perpendicular field function, H_y, and the tape magnetization I_y'' from an analogous reciprocity equation to eq. (4.4)

$$\varphi_c = \frac{2\pi}{U_0} \int_{1-(a+c)}^{1-a} dy \int_{-L}^{+L} H_y I_y'' \, dx \, . \tag{4.27}$$

It has been shown that for a well shielded head the magnetization of the tape in contact with the head in zone A provides the greater part of the reproducing head core flux. In this case it is found that the head flux is proportional to

$$L \sin \frac{2\pi L}{\lambda} \Big/ \left(\frac{2\pi L}{\lambda}\right)$$

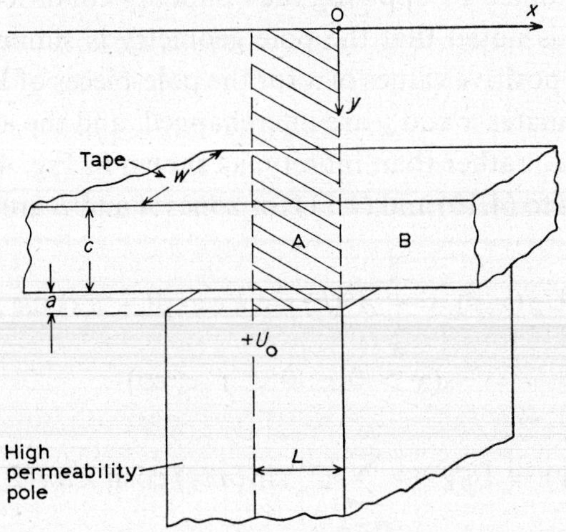

Fig. 4.9.   Reproduction with single contacting pole head.

and thus has an analogous 'gap' loss associated with the pole width $2L$. However, as the recorded wavelength approaches infinity, the gap loss factor approaches L, and the head core flux therefore approaches a finite value. This is only true in the case of a ring head when the pole-pieces become infinitely long. For finite pole-piece length in a ring head the core flux approaches zero as $\lambda \to \infty$.

It is also finally noted that, as in the case of longitudinal recording and reproducing with a ring head, the core flux is in phase with the tape magnetization.

§ 4. REPRODUCTION OF PULSE RECORDING WITH RING HEAD

The analysis of the process of reproducing recorded signals having binary rather than sinusoidal waveforms can be accomplished by application of the principle of reciprocity, as already described. In the most common form of recorded binary information (NRZ) the tape is saturated at all times and the direction of magnetization reverses in accordance with the information to be recorded. Another form of recording, (RZ), may be mentioned in which short pulses of saturated tape occur in accordance with the recorded signal: in this case the tape is unmagnetized in the absence of recorded information. A description of the recording process for these two techniques may be found in Chapter 3, § 1.2.

The reciprocity equation for the reproduced voltage from a pulse recording includes terms which are related to the reproducing process, i.e. $H(x, y)$, and terms due to the recording process, i.e. $I''(x, y)$. In Chapter 3, § 1.2 recording effects have been emphasized by assuming that the reproducing process has a relatively high resolution. In this case the reproducing head field function may be approximated by an impulse and the reproduced voltage becomes the derivative of the longitudinal tape magnetization along the tape. Here it is desired to emphasize the reproducing process and this is done by assuming that the recording process has a relatively high resolution. In this case, for instance, RZ recording may be approximated by an impulse, and NRZ by an infinitely short step for the change of magnetization polarity. Practical conditions appear to lie somewhere between these two extremes with the reproducing process having somewhat lower resolution (TEER [1961], ELDRIDGE [1960]).

Since, according to the reciprocity equation, the flux in the reproducing head core is determined by multiplying the reproducing head field function $H(x, y)$ by the tape magnetization function $I''(x, y)$, the core flux is proportional to the head field function when a single ideal RZ pulse is transported over the head gap. This is because, at any instant, the integration of the product HI'' over the length of the tape is proportional to the field, H, at the position of the pulse. In the case of a single step from $I'' = 0$ to $I'' = I''_{max}$ the instantaneous reproducing head core flux is proportional to the integration of the head field function up to the position of the flux step. The reproducing head voltage is then proportional to the field function at the position of the step. In NRZ recording the magnetization steps correspond to changes from $- I''_{\max}$ to $I''_{\max}$ as shown in Fig. 4.10. The reproducing head core flux will be first calculated for an instantaneous magnetization reversal and then the practical condition of finite magnetization reversal lengths is considered.

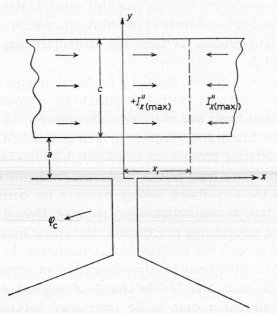

Fig. 4.10.   Reproduction of ideal unit step function located at $x = x_1$.

§ 4.1.   *Reproduction of Ideal Step Function*

This simple and readily analysable situation can be employed to give useful information about the direction of magnetization in a tape and the field configuration around a head. The reproducing head response to a recorded step function is in the form of a unidirectional pulse. From the point of view of high density recording of NRZ information, which amounts to a coded sequence of such step functions, it is important to know how the reproduced pulse shape and height depend on such parameters as coating thickness, head-to-tape spacing, and gap length.

The desirable form of the reproduced pulse is one of large amplitude and small width, and these characteristics are therefore chosen for evaluation of the reproduction process. It will also become apparent how the recording process affects the final reproduced pulse characteristics.

Derivation of Reproducing Head Output

Considering both longitudinal and perpendicular components in the recorded magnetization, the head core flux from an infinitely thin layer is given by

$$\varphi_c = \frac{2\pi}{U_0} \int_a^{a+c} dy \left[\int_{-\infty}^{+\infty} H_x I_x'' \, dx + \int_{-\infty}^{+\infty} H_y I_y'' \, dx \right], \qquad (4.28)$$

the symbols being as indicated in eq. (4.4).

Considering, initially, the longitudinal component of an ideal recorded step function $I_{x(\max)}''$ as illustrated in Fig. 4.10.

$$\int_{-\infty}^{+\infty} H_x I_x'' \, dx = 2 I_{x(\max)}'' \int_0^{x_1} H_x \, dx . \qquad (4.29)$$

Now, the reproducing head voltage, e_x, due to the longitudinal component of magnetization is given by the time derivative of the core flux. Thus,

$$e_x = - n(d\varphi_c/dt) = - nv(d\varphi_c/dx_1) \qquad (4.30)$$

where $v = dx_1/dt$, the tape velocity, and n is the number of turns on the core. Combining this with eqs. (4.28) and (4.29)

$$e_x = \frac{4\pi nv I_{x(\max)}''}{U_0} \int_a^{a+c} H_x \, dy , \qquad (4.31)$$

and, by similar reasoning,

$$e_y = \frac{4\pi nv I_{y(\max)}''}{U_0} \int_a^{a+c} H_y \, dy . \qquad (4.32)$$

Thus, for a constant tape velocity, the reproduced voltage waveform from an ideal step function is proportional to the corresponding reproducing head field contour for any elemental thickness of the coating. It is, therefore, possible to investigate the effects of gap length, head-to-tape spacing and tape thickness on the reproduced waveform for an ideal step function.

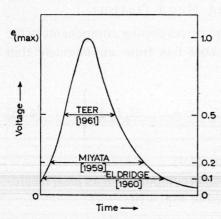

Fig. 4.11. Typical reproduced pulse shape for ideal recorded step function. References to various pulse width definitions.

Spacing

Considering distances from the head for which $y > g'$, the field distribution around the head closely approximates that from an infinitely small gap; hence

$$H_x = \quad 4ni[y/(x^2 + y^2)].$$
$$H_y = - \; 4ni[x/(x^2 + y^2)].$$

(4.33)

Thus the reproducing head voltages for longitudinal and perpendicular magnetization step functions are respectively

$$e_x/v = \frac{8\pi n^2 i}{U_0} I''_{x(\text{max})} \ln \left[\frac{x^2 + (a + c)^2}{x^2 + a^2} \right],$$

(4.34)

$$e_y/v = \frac{16\pi n^2 i}{U_0} I''_{y(\text{max})} \left[\tan^{-1}\left(\frac{a}{x}\right) - \tan^{-1}\left(\frac{a + c}{x}\right) \right].$$

(4.35)

The reproduced pulse shape for a recorded step function is shown in Fig. 4.11 which shows approximate symmetry about the maximum value. This pulse shape is in very approximate correspondence with eq.

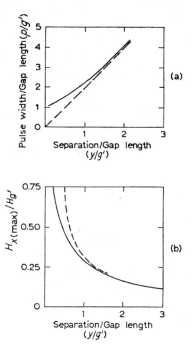

Fig. 4.12.   a. Reproduced pulse width (p) as a function of spacing (y).
b. Reproduced pulse height ($e_{x\,(\text{max})} \propto H_{x\,(\text{max})}$) as a function of spacing (TEER [1961]).
————— Derived from accurate field function.
— — — Derived from $g' = 0$ field function.

(4.34) for a longitudinally recorded step function. The observed asymmetry could be explained by assuming a perpendicular magnetization component. However, it has been shown, by comparing the practical reproduced pulses with those from purely longitudinal magnetization step functions, that the asymmetry is similar in the two cases and is a feature of the recording process (see Ch. 3, § 1.2). Perpendicular magnetization is therefore neglected in the analysis of the reproducing function (ELDRIDGE [1960], TEER [1961]). The simple field formulae of eqs. (4.34), (4.35) are not accurate for $y < g'$. In this case it is necessary to use the field function equation derived in Section 2.3 (eq. (4.16)). By applying a similar field function, obtained by the method of conformal mapping (TEER [1961]), the pulse width and height may be computed as a function of head-to-tape spacing. Here the pulse width is defined as the time between instantaneous amplitudes of $0.5e_{(max)}$. Fig. 4.12a and b, show the pulse width and height respec-

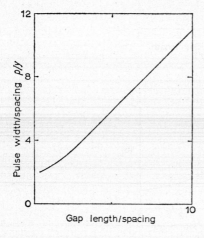

Fig. 4.13.   Relative pulse width as a function of reproducing head gap length.

tively obtained in the above manner, assuming a tape thickness approaching zero. The dotted lines indicate the relationship obtained using the zero gap length formula for H_x (eq. (4.33)), showing accurate representation of pulse height for $y > g'$ and of pulse width for $y > 2g'$.

Coating Thickness

The effect of coating thickness on pulse width and height may be assessed by appropriate summation of the elemental layers using the curves in Fig. 4.12. As may be expected, an increase of coating thickness gives less than a corresponding increase of the reproduced pulse amplitude along with an increase of the pulse width.

Gap Length

Fig. 4.12a, showing the reproduced pulse width (*p*) as a function of spacing from the head, may be redrawn to indicate its relative dependence on gap length as shown in Fig. 4.13. Again, this curve refers to a coating thickness approaching zero, but good agreement is obtained

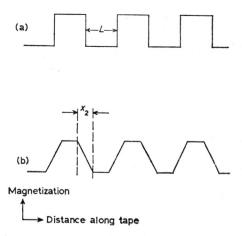

Fig. 4.14.   Tape magnetization patterns assumed for reproduction of NRZ recordings.
a. Ideal waveform.
b. Trapezoidal waveform.

in practice if the recorded depth is small, thus minimizing the effects of finite coating thickness and self demagnetization (TEER [1961]).

The reproduction of recorded pulses as obtained in practice involves taking account of all of the parameters considered here, that is spacing, coating thickness, gap length, plus the parameters controlling the recorded pulse including recording head gap length, coating thickness, spacing and demagnetization effects. The parameters affecting the recording are considered in Chapter 3, § 1.2. Reproduction of such recordings is discussed in the following section.

§ 4.2.   *Reproduction of Practical NRZ Recordings*

It has been shown in the description of pulse recording (Ch. 3, § 1.2) that the recorded magnetization change in NRZ recording differs significantly from the ideal step function, considered in the previous section, especially when relatively thick coatings are used. In such cases both the rate of change of the magnetization reversal and its relative longitudinal position are progressively changed at different depths in the coating, and the resultant magnetization pattern for a series of NRZ pulses is not a simple waveform. A demonstration of the effects of practical recorded waveforms may be obtained by considering a trapezoidal approximation as shown in Fig. 4.14b rather than the ideal step reversals of Fig. 4.14a. Here a 'bit' length, L, is depicted with constant rate of magnetization reversal taking place over a finite length x_2. The trapezoidal magnetization waveform approximation may then be analysed into a Fourier series and the reproduction of the sinusoidal components may be considered as previously described (SCHOOLS [1961], KOSTYSHYN [1962]). The resultant summation shows that the reproduced pulse width, defined as indicated in Fig. 4.11, and the pulse amplitudes increase with coating thickness in a similar manner to that described for the ideal magnetization reversal. However, in contrast to the linear dependence of pulse width on the reproducing head gap length, shown previously, little change is obtained for the trapezoidal model. These results are generally in good agreement with measured results for similar bit densities (MIYATA and HARTEL [1959], ELDRIDGE [1960]). Better recording resolution in the future can be expected from tapes having thin coatings and high ratios of H_c/I_r'' which will reduce coating thickness effects and self demagnetization effects. As the ideal recorded step function is approached, the reduction of pulse width with reproducing head gap length may be used with advantage. Further significant reduction in the reproduced pulse width may be obtained by appropriate equalization techniques, particularly in reproduction (FAN [1961b]).

Finally, the reproduction pulse height and width may be calculated for trapezoidal magnetization waveforms for a range of packing densities. As the number of reversals per unit length of tape is increased above about 200 bits per inch, losses occur both in recording and reproduction and the output approaches zero at about 5 000 bits

per inch when the transition length and the tape thickness are assumed
to be 0.5 mil. Recording losses are due to tape elements experiencing
fields of opposite polarities during their passage through the recording
field, as described for sinusoidal recording without bias (Ch. 3, § 1.1).
Reproduction losses are associated with the thickness and spacing
losses due to the short wavelength components of the recorded waveform.

§ 5. REPRODUCING HEAD DESIGN CONSIDERATIONS

In practice, the reproducing head core flux associated with the
voltage developed in the head coil is usually less than that calculated
in the previous section. This calculated flux is obtained by taking into
account losses inherent to the reproducing process with a gapped ring
head. Reproducing heads of good design can, however, be made with
such performance that the calculated reproduction losses associated
with the tape are the dominant losses; that is to say, the present state
of high resolution reproduction techniques is such that the major
limitations occurring are due to the so-called thickness and spacing
losses. Although the spacing loss is a function of the combined head
and tape roughnesses for 'in-contact' reproduction, it is the latter which
limits the contact. On the other hand, careful adjustment of the geometry of the recorded track and the reproducing head core is required
to achieve an optimum practical efficiency and the possible contributing practical losses will now be considered.

§ 5.1. *Efficiency of Head Magnetic Circuit*

Contour Effect

The design objective for the reproducing head core is to achieve a
maximum instantaneous core flux proportional to the surface induction of the tape. As described earlier in this chapter, it is important to
limit the reproducing head gap length, g', to be less than half the
shortest recorded wavelength in order to avoid significant interference
losses in the gap region. In addition to this, what amounts to a secondary gap effect must be considered; this is due to the finite length of

contact, G, between the pole-pieces and the tape as illustrated in Fig. 4.15. Here, possible flux paths external to the tape are drawn for the condition that the recorded half wavelength is somewhat longer than the contact length G. Although some of the external flux can link with the reproducing head coil outside the head-to-tape contact length, as

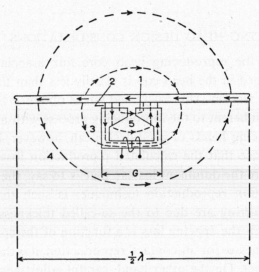

Fig. 4.15.   Reproduction of long wavelengths by ring head.
1 – 5: possible flux paths.

illustrated by path (3) in Fig. 4.15, this is normally attenuated by a magnetic shield around the head. Flux paths starting remote from the head, such as (4), do not link with the core at all. Consequently, the flux enters the head core only along the contact length G, which limits the flux to a fraction, $2G/\lambda$, of the total external flux.

The flux configuration shown in Fig. 4.15 is, however, too simple to portray adequately the head core flux pattern when the length of the head is not large compared to the recorded wavelength. In the derivation of the head core flux (§ 2) it was assumed that the length of the pole-pieces was infinite when calculating the effective head field distribution, H_x. For finite pole-piece lengths, the field distribution in the gap region will be similar to that previously calculated, but there will be secondary field maxima near the ends of the pole-pieces. These are of opposite sign to the gap field as shown in Fig. 4.16. Sinusoidal

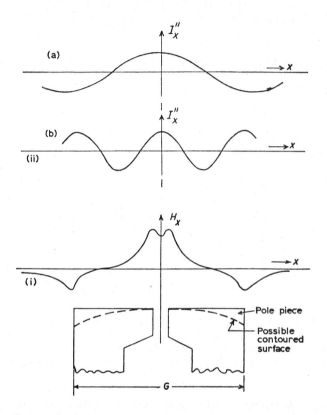

Fig. 4.16.   (i) Longitudinal field distribution, H_x, near reproducing head.
(ii) Recorded magnetization waveforms: (a) $G/\lambda = 0.9$; (b) $G/\lambda = 2.0$.

recorded waveforms, I''_x, are also shown in Fig. 4.16 corresponding to $G/\lambda = 0.9$ and 2.0.

The head core flux is given by the integral $\int_{-\infty}^{+\infty} H_x I''_x$ (eq. (4.4)). In the first case, the contributions of the recorded flux near the extremities of the poles aid that from the gap region, whereas in the second case they oppose it. The head core flux, φ'_c, will therefore have maximum and minimum values as the wavelength changes with respect to the pole-piece length G. For a head of the type depicted in Fig. 4.16, if G is not negligible with respect to λ,

$$\varphi'_c = \varphi_c \left\{ 1 - \frac{0.205 \cos \left[\pi - (G/\lambda + \frac{1}{6}) \right]}{(G/\lambda)^{\frac{2}{3}}} \right\}. \qquad (4.36)$$

Here φ_c is the core flux for infinite pole length as given in eq. (4.19).

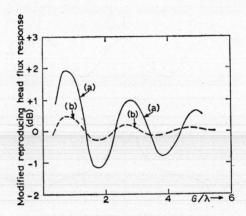

Fig. 4.17.   Calculated relative reproducing head core flux for finite pole length, G, as shown in Fig. 4.15.
Curve (a) — tape contacting head completely.
Curve (b) — tape contacting head near gap only (WESTMIJZE [1953a]).

The above approximate expression was obtained by calculating the head field contour using conformal mapping (WESTMIJZE [1953a]), assuming the head and tape permeabilities to be infinite and unity respectively. It results in an undulating response as shown in Fig. 4.17 – curve (a). These unwanted undulations can be attenuated by introducing contoured pole-pieces, giving an extra separation loss term which increases with longitudinal distance from the gap. An example of such contouring is shown by the dotted line in Fig. 4.16. In this case the tape would not leave the head at a sharp angle and so the longitudinal field along the tape path would not have the sharp secondary maxima. Thus the undulating response is attenuated as shown in curve (b) (Fig. 4.17). Other techniques for reducing the coherence of the secondary gap effect include making the pole-piece lengths different on each side of the gap (MANKIN [1952]).

For recorded wavelengths less than $\lambda \approx G/5$, the undulating response disappears since the recorded wavelengths begin to interfere in the relatively broad secondary gap response zone; hence the secondary gap loss is high. At the other extreme, as the recorded wavelength becomes infinitely long, the integral $\int_{-\infty}^{+\infty} H_x I_x'' \mathrm{d}x$ becomes zero and hence the reproducing head flux falls to zero.

Core Efficiency

Having discussed the reproducing head design from the point of view of maximizing the core flux relative to the tape surface induction, it remains to consider how the maximum part of the flux can be induced to couple with the reproducing head winding. As can be seen in Fig. 4.15, the pole-piece flux can take path (1) across the gap or path (2) around the core. Certain practical considerations dictate that the gap reluctance be quite small; this is due primarily to the requirement that the gap length, g', be less than the shortest recorded half wavelength in order to avoid gap loss effects. There is also a lower limit to a practical gap depth, due to the subsequent wearing of a head by the tape. Hence the reluctance of flux path (1) is necessarily relatively low due to the low gap reluctance, $R_{g'}$. High efficiency, therefore, depends on making the head core reluctance R_c, small with respect to $R_{g'}$. The fraction of head core flux passing through the desired path (2) is given by, $\varphi_c'(2)$, where

$$\varphi_c'(2) = \frac{R_{g'}\varphi_c'}{R_c + R_{g'}} = \frac{(g'/A_{g'})\varphi_c'}{(g'/A_{g'}) + (h/\mu_c A_h) + (i/A_i)}, \quad (4.37)$$

where g' and i are the front and rear gap lengths, and h is the core length; further $A_{g'}$, A_h, A_i are the corresponding cross sectional areas and μ_c is the core permeability. Some air gap in the core is inevitable in any practical design. Since the most common core design consists of two half cores, there will be a gap in the core circuit in the same plane

TABLE 4.2

System	Top frequency f	Tape speed v (inch/sec)	Gap area / Core area (A_g'/A_h)	Permeability (μ_c)	Core length (h) (inch)	Gap length (g') (mil)	Efficiency $\dfrac{1}{1 + \dfrac{h}{g'}\dfrac{A_g'}{A_h}\dfrac{1}{\mu_c}}$
Professional audio recording	15 kc/s	15	0.4	20 000	2	0.5	0.93
High resolution audio recording	15 kc/s	$1\tfrac{7}{8}$	0.1	20 000	1	0.05	0.91
Direct video recording	3 Mc/s	200	0.1	1 000	1	0.03	0.23
Direct video recording	3 Mc/s	200	0.1	20 000	1	0.03	0.86

as the pole-piece gap; this is referred to above as the rear gap. Fortunately, the rear gap reluctance can be made as small as desired by increasing the cross sectional area of the core. If, on the other hand, the reluctance of the core is greater than that of the gap, less than half the pole-piece flux will be linked with the pickup coil. Some advantage may be obtained in this case by winding the pickup coil nearer to the pole-pieces since significant flux leakage occurs, which by-passes the rear leg of the core as illustrated by path (5) in Fig. 4.15.

Apart from the obvious step of keeping the length of the core as short as possible, compatible with obtaining a sufficient number of turns in the coils, the remaining parameter controlling the useful core flux, $\varphi'_c(2)$, is the core initial permeability, μ_c. It was pointed out in § 2.5 that the pole-piece flux φ'_c is not greatly changed for working permeabilities above 100 or so. However, such low permeabilities can have an important effect on the division of the pole-piece flux between the front and back gap circuits. This is illustrated in the practical examples quoted in Table 4.2, where high permeability nickel-iron alloy laminations ($\mu_c = 20\,000$) are used at audio frequencies, and moderate permeability ferrites ($\mu_c = 1000$) at high frequencies. For comparison, the efficiency is also calculated for alloy laminations at 3 Mc/s, although, as will be seen later, the practical effective permeability is much reduced. It is assumed that the reluctance of the rear gap can be ignored for the designs considered.

From the results in Table 4.2 it can be seen that the small gap length necessary for high resolution audio recording can be offset by a reduction of the gap depth. This is rather more risky for the high speeds used for direct video recording or pulse recording due to the increased wear of the head. Even if this is done, the high efficiency is not maintained if ferrite cores are used. Nevertheless, ferrites are more efficient than high permeability alloys for the high frequencies occurring in video and pulse recording due to their inherently large resistivity which minimizes eddy current losses (see following section).

Until recently a compromise core construction has been commonly used for video frequency recording and reproduction comprising a ferrite core and a hard high permeability aluminium-iron alloy called

Alfenol (Kornei [1956]). The prime advantage of this alloy is its good wearing properties; it does, however, contribute the major reluctance to the core circuit at high frequencies and therefore controls the head sensitivity (Frost [1960]). More recently, significant improvements have been made in techniques for producing durable high-resolution ferrite pole-pieces (Duinker [1960]), allowing the advantage of all-ferrite cores to be exploited. It is claimed that such cores are not only optimum for video recording but also have advantages for slow speed high-resolution audio recording (see Ch. 7). This is due, in part, to the maintenance of high permeability in the surface layers of the polished pole-pieces, whereas polished metal pole-pieces form a low permeability skin which is not removed on subsequent annealing.

§ 5.2. *Practical Reproduction Losses*

In addition to the losses inherent in the reproducing process using a gapped ring head, several other sources of imperfection can lead to further losses; these will be briefly mentioned here in order to illustrate design features which might be included to minimize their severity.

Practical Gap Losses

Imperfections of the reproducing head core have mechanical and magnetic origins. Mechanical imperfections in the front gap can significantly deteriorate the short wavelength resolution of the head. For instance, if the gap corners are rounded off then a kind of separation loss occurs for short wavelengths. This is analogous to an increase in the gap loss which lengthens the effective gap and reduces the short wavelength response. For example, if the rounded gap edges correspond to a radius of 0.3 micron for a mechanical gap length of $g' = 1$ micron, the effective length of g' increases from $1.12g'$ to $1.38g'$; furthermore, the response to a recorded wavelength of $\lambda = 1.6g'$ is reduced by 6.8 dB compared to that for perfectly sharp edges (Duinker [1961]). Gap edge rounding can be mechanical or magnetic in origin. The polishing procedure used on metal laminations tends to produce a reduced permeability skin, giving an effective separation loss and gap edge rounding. Ferrite gaps have been made without significant edge rounding due to either mechanical or magnetic defects and may be intrinsically superior to metal laminations for high resolutions (DUINKER [1961]).

In the foregoing analyses of the reproduction process it has been tacitly assumed that the flux in the reproducing head can be obtained by multiplying the calculated flux per unit gap width by the physical gap width, w. However, if the reproducing head gap edges are not both parallel to the trailing edge of the recording head, the flux per unit width may vary in phase along the width of the head, leading to loss of resolution. In effect, the reproducing head then picks up an average of the recorded flux over some finite length of the recording determined by the angle, ϕ, between the reproducing head gap and the recording. Thus the effect is similar to the gap loss described earlier. The reproducing head voltage e_ϕ under these conditions is given approximately by (DANIEL and AXON [1953]).

$$\frac{e_\phi}{e_0} = \frac{\sin(\pi w \phi / \lambda)}{\pi w \phi / \lambda} \, . \tag{4.38}$$

Modifications of this misalignment loss can also occur if the reproducing head gap edges are not parallel to each other or if the edges are not straight. Misalignment losses may be reduced to an acceptable level by restricting the track width when very short wavelengths are to be reproduced, and by taking care to produce straight recording and reproducing head gaps. Of course, lower signal-to-noise ratios are obtained from narrower track widths, a reduction of 3 dB being obtained for each halving of the width. Some idea of the degree of mechanical perfection obtained in high resolution reproducing head gaps may be obtained from Fig. 4.18b and c which show gaps of the order of one micron in length in high permeability alloy poles and ferrite poles respectively. These results may be obtained with metal shims or deposited gap materials. Unless careful polishing techniques are used, the alloy pole edges can become ragged due to smearing of material across the gap as shown in Fig. 4.18a. Of course, there is a tendency for such smearing to occur in normal abrasion by the tape and, where high tape speeds are used, it is advantageous to use a hard alloy like Alfenol or ferrite. Ferrites are brittle materials and gap edge wear takes the form of chipping off small grains. Such gap edge erosion has however been virtually eliminated by bonding a suitable glass to the ferrite, the glass then acting as the gap spacer (DUINKER [1960]).

Practical Core Losses

As already indicated, the reluctance of the reproducing head core must be held low with respect to the gap reluctance in order to obtain efficient flux coupling to the head coil. At high frequencies, however, the effective permeability of the core is reduced due to eddy current losses, hysteresis losses, and residual losses caused by magnetic viscosity effects. The net effect of these losses is to cause a reduction of the core flux, which is confined to the periphery of the core, and a time lag between the flux and the applied field. Thus, the permeability of the core at a frequency, f, is complex and may be defined in terms of real and imaginary components μ'_c and μ''_c respectively where μ'_c may be regarded as the magnetization component and μ''_c the loss component. Hence,

$$\mu_f = \mu'_c - i\mu''_c, \tag{4.39}$$

or, in terms of the loss angle, θ,

$$\mu_f = \mu_c(\cos \theta - i \sin \theta), \tag{4.40}$$

where μ_c is the modulus of the complex permeability. The loss factor, $\tan \theta$, is then equal to μ''_c/μ'_c.

Two types of core material are in common use for reproducing heads, laminated alloy cores and ferrites. For the first, the chief magnetic limitation arises from eddy current losses occurring in the relatively low resistivity alloys (resistivity $\approx$ 50 μ ohm cm.). The resulting components of the complex permeability are approximately given by (CAUER [1925])

$$\mu'_c = \frac{\mu_c}{\alpha} \frac{\sinh \alpha + \sin \alpha}{\cosh \alpha - \cos \alpha}$$

$$\tag{4.41}$$

$$\mu''_c = \frac{\mu_c}{\alpha} \frac{\sinh \alpha - \sin \alpha}{\cosh \alpha + \cos \alpha}$$

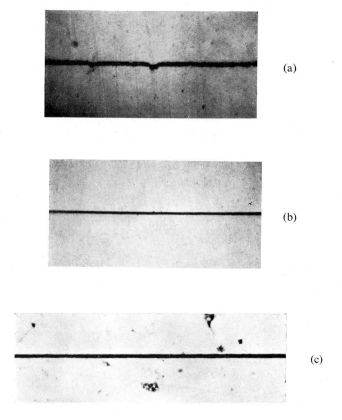

(a)

(b)

(c)

Fig. 4.18.   High resolution reproducing head gaps in high permeability poles:
a. Gap length $\approx 1\ \mu$. Coarse polish. Alloy poles.
b. Gap length $\approx 1\ \mu$. Fine polish. Alloy poles.
c. Gap length $= 0.7\ \mu$. Ferrite poles (DUINKER [1960]).

where μ_c is the zero frequency permeability, and $\alpha = 2\pi\delta(\mu_c f/\rho_c)^{\frac{1}{2}}$. Thus α depends on the lamination thickness δ, the material resistivity, ρ_c, and the applied frequency, f. For audio and video reproducing heads laminated alloys have been used with respective typical thicknesses of 4 and 2 mil. Assuming $\mu_c = 20\,000$ and upper frequencies of 15 kc/s for audio and 3 Mc/s for video, α has the values of 5 and 35 respectively. In both cases, eq. (4.41) reduces to

$$\mu'_c \approx \mu''_c \approx \mu_c/\alpha . \qquad (4.42)$$

Hence, the magnetization component of the effective permeability is considerably reduced for both cases. The effect on the head sensitivity may be assessed (DANIEL et al. [1957]), by calculating the change of core flux as the frequency is raised by applying eq. (4.37). The ratio of desired core flux $\Phi_f = \varphi'_c(2)$ at the operating frequency to that at low frequencies Φ_c is then given by

$$\frac{\Phi_f}{\Phi_c} = \frac{\dfrac{g'}{h}\dfrac{A_h}{A_{g'}} + \dfrac{1}{\mu_c}}{\dfrac{g'}{h}\dfrac{A_h}{A_{g'}} + \dfrac{1}{\mu_f}} . \qquad (4.43)$$

Here it is assumed that the rear gap $i = 0$. Substituting for μ_f from eq. (4.39) gives

$$\frac{\Phi_f}{\Phi_c} = \frac{\gamma\mu_c + 1}{\mu_c}\left[\frac{(\mu'_c)^2 + (\mu''_c)^2}{(\gamma\mu'_c + 1)^2 + (\gamma\mu''_c)^2}\right]^{\frac{1}{2}} \qquad (4.44)$$

where

$$\gamma = \frac{g'}{h}\frac{A_h}{A_{g'}} .$$

In the case of alloy laminations, eq. (4.41) is applicable and hence

$$\frac{\Phi_f}{\Phi_c} = \frac{(\gamma\mu_c + 1)(2^{\frac{1}{2}})}{[(\gamma\mu_c + \alpha)^2 + (\gamma\mu_c)^2]^{\frac{1}{2}}} . \qquad (4.45)$$

Applying this equation to the audio reproducing head geometries considered in the first row of Table 4.2, using 4 mil laminations, the sensitivity at 15 kc/s is 90% of the low frequency sensitivity. On the other hand, the video reproducing head sensitivity at 3 Mc/s using 2 mil laminations is reduced to only 26% of its low frequency value. Using the reproducing head efficiency figures calculated in Table 4.2, the practical high frequency efficiencies for audio and video recording are 0.84 and 0.22 respectively. Thus it would appear that, at low frequencies, high efficiencies can be maintained in laminated alloy

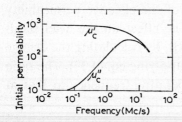

Fig. 4.19. Real and imaginary permeability for nickel-zinc ferrite ($Ni_{0.26} Zn_{0.64} Fe_2 O_4$) (DUINKER [1960]).

cores, but for applications in the megacycle range ferrite cores may become preferable.

The real and imaginary component μ_c' and μ_c'' of the complex initial permeability for a nickel zinc ferrite, of the type used for recording and reproducing heads, are plotted in Fig. 4.19 as a function of frequency (DUINKER [1960]). The rapid rise of μ_c'' is due to electron spin resonance effects in the ferrite. In general, ferrites with high values of μ_c' have low cut-off frequencies due to low resonance frequencies. This can be readily understood since the resonance frequency depends on the magnitude of the internal fields responsible for crystal anisotropy; when these are low then the initial permeability is high. For the material illustrated in Fig. 4.19, at 3 Mc/s $\mu_c' = 700$ and $\mu_c'' = 300$. Substituting in eq. (4.44) gives $\Phi_f/\Phi_c = 0.82$, showing high efficiency comparable to the low frequency value. As shown in Table 4.2, the efficiency of low permeability ferrite heads is poor, but the lack of high frequency losses offsets this deficiency at high frequencies, making ferrites preferable. In fact, the efficiency of ferrite cores is probably greater than thin lamination alloy cores at somewhat lower frequencies since the effective low frequency permeability of the latter is probably less than 20 000 for ultra thin laminations. On the other hand, the low losses in ferrite cores imply low intrinsic head noise (DUINKER [1960]). Future developments in soft ferrites can be expected in which higher initial permeabilities will be obtained extending their application in reproducing heads.

§ 6. REPRODUCTION NOISE

When a recording is reproduced with a conventional head in the manner described in this chapter, the voltage obtained at the reproducing head coil is due to the time rate of change of the tape magnetization in the vicinity of the reproducing head gap. The tape magnetization in each elementary tape volume is a combination of unidirectionally magnetized tape particles carrying the required signal information and particles whose magnetizations are directed at random, since they were not affected by the signal field during recording. The time rate of change of the tape magnetization may be looked upon as being the product of the spatial rate of change of magnetization and the relative velocity between the tape and the reproducing head. The reproduced signal voltage then arises from the interaction between tape particles, magnetized by the recording field, and the reproducing head when the tape is transported with constant velocity in the vicinity of the head gap. Any additional interaction between the tape magnetization and the head gives rise to unwanted reproduced voltages which are categorized as reproduction noise. Layer to layer print-through recording comes under this general heading and is described separately in § 7.

There are many possible sources of reproduction noise, the most fundamental being that detected by the reproducing head from particles whose magnetization has not been reoriented by the recording signal. These particles are magnetized in random directions and produce a small random flux in the reproducing head core, thus setting a lower limit to the achievable noise level. This process is described later and it will be shown that the random flux in the reproducing head core increases with the magnitude of the tape magnetization. At first sight it appears contradictory that, as particles become locally unidirectionally magnetized by a recording signal, the randomly varying part of the interaction with the reproducing head should also increase. This is due, however, to departures from the ideal model of tape particles as being identical, uniformly dispersed, non-interacting single domains.

In addition to departures from the spatial uniformity of particles in a tape there are also deviations from uniform transportation of the

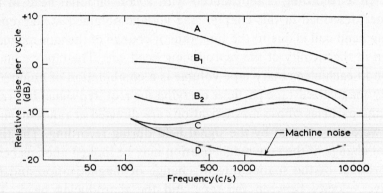

Fig. 4.20.   Noise spectra for various erasure methods.
Tape speed 15 inch/sec.
B_1 — 100 kc/s bias B_2 — 600 kc/s bias. (VON BEHREN [1956]).

particles past the reproducing head gap; these can also give rise to unwanted reproduced voltages, or noise. For instance, when tape is pulled past the reproducing head, it is very difficult to avoid exciting small longitudinal vibrations which cause frequency modulation of the recorded signal. Unwanted perpendicular motion of the tape can also occur due to foreign particles or surface nodules on the tape which lift the tape momentarily out of contact with the reproducing head pole-pieces. On a recorded tape such motion leads to rapid amplitude changes in the reproducing head flux and consequent noise voltage pulses. Finally, it should be borne in mind that noise can also originate in the reproducing head core and in the reproducing amplifier electronic circuitry. However, the following discussion is restricted to noise which is a function of tape characteristics.

§ 6.1.　*Background Noise*

The random noise originating from the tape during reproduction may be conveniently classified into two types, background noise and modulation noise. The background noise is the result of reproducing demagnetized tape, and the modulation noise is the added noise which is a function of the tape magnetization level.

In studying the background noise it is first necessary to consider the efficiency of the demagnetization mechanism in removing any previous magnetization. Different methods for achieving demagnetization produce different noise levels, as illustrated in Fig. 4.20, where the noise spectra are plotted for various erasing conditions used in magnetic recording. Curves B_1 and B_2 correspond to erasure using a wide gap head supplied with an ac signal having a minimum of even-harmonic components; subsequently the tape is run over a recording head carrying ac bias but no recording signal. The relatively high noise levels are due to an insufficient number of magnetization cycles during the time of transit across the recording head zone. Lower noise spectra, curve C, are obtained when the tape is bulk erased. Bulk erasure consists of subjecting the tape to a high amplitude alternating field which is spread over a relatively large volume, giving a slow field decrement on withdrawing the tape from the field. The mechanism of tape erasure is similar to anhysteretic magnetization in zero external dc field. Even harmonic components in the erasing field produce a dc field component leading to a macroscopic remanent magnetization in the tape. The considerably higher noise level due to dc magnetization is shown in curve A, Fig. 4.20. It has the same origin as modulation noise, due to signal magnetization, which is described in the next section. In the absence of dc components in the erasing field, the tape particles become anhysteretically magnetized in the particle interaction fields, resulting in random magnetization if the interaction fields are random. To achieve true anhysteretic magnetization in the local interaction field, it is necessary to ensure that the external ac field has a sufficiently large maximum amplitude to switch the magnetization directions of all particles. It is further necessary that many cycles occur during the field reduction, and that the field is sufficiently extensive to switch all particles in the vicinity of those being erased. These conditions are not always met in practical erasing and ac biassing techniques.

Assuming that curve C of Fig. 4.20 represents the noise at the reproducing head due to a completely demagnetized tape, some idea of the spectral distribution of noise sources in the tape may be obtained by correcting the curve for the reproduction losses. Assuming that the reproduction losses for noise are similar to those for sinusoidal signals in the same frequency band and that no recording losses occur, then the constant current recording characteristic may be used to correct the reproducing head noise spectrum. The resulting spectrum of noise sources is approximately flat at high frequencies and gradually increases with decreasing frequencies. However, low frequency noise measurements are difficult to perform due to the extremely small reproduced voltage, and may also be inaccurate due to the possible occurrence of residual long wavelength magnetization components. As will be seen later, long wavelength noise components increase with the tape magnetization level, and may be attributed to departures from the ideal tape coating consisting of a uniform distribution of independent and identical particles. A flat spectrum of noise sources, called white noise, can be accounted for by the particulate nature of such an ideal tape coating, and this is the fundamental nature of the background noise.

Basically the reproduction processes for signal and noise are the same, that is to say, the mechanism of detection of particle magnetization is the same for randomly or unidirectionally magnetized particles. However, in the case of the randomly magnetized particles, the surface tape fields are produced by the statistical addition of the fields from the particles in the tape layer. Assuming, for simplicity, that the particle magnetic moments ($\pm m$) are oriented in the longitudinal direction of the tape, and have equal probability of magnetization in either direction for a demagnetized tape, the reproducing head instantaneous core flux (φ_c) is given by

$$\varphi_c = m \, \Sigma \, \varphi_p, \qquad (4.46)$$

where φ_p is the core flux for particle of unit moment. Similarly, the mean core flux ($\bar{\varphi}_c$) is given by

$$\bar{\varphi}_c = m \, \Sigma \, \bar{\varphi}_p. \qquad (4.47)$$

This is zero for a demagnetized tape when the mean is taken over many particles. The noise core flux for a discrete number of randomly magnetized particles is defined in terms of the mean square deviation from the mean flux (φ_n^2), or the variance. For the ideal tape with N identical independent particles contributing to the output

$$\varphi_n^2 = \bar{\varphi}_c^2 - (\bar{\varphi}_c)^2.$$
$$= m^2 N \{\bar{\varphi}_p^2 - (\bar{\varphi}_p)^2.\} \qquad (4.48)$$
$$= m^2 Dw \int_a^{a+c} dy \int_{-\infty}^{+\infty} \varphi_p^2 \, dx \,,$$

where D is the particle density. The other symbols refer to a coordinate system similar to Fig. 4.3. and w is the tape width. Eq. (4.48) is the noise reproduction analogy of eq. (4.4) for signal reproduction.

The evaluation of the reproducing head noise flux can then be accomplished by determining the integral of the head sensitivity over the volume of the tape as described in § 2. A simple single turn head with an infinitely small gap is considered here; in this case

$$\varphi_p = 4y / (x^2 + y^2) \qquad (4.49)$$

which is derived by reciprocity from eq. (2.2). The mean square noise core flux and noise voltage are then

$$\varphi_n^2 = 8\pi Dm^2 w \ln \{(a + c)/a\} \,, \qquad (4.50)$$

$$e_n^2 = 4\pi Dm^2 wv^2 c(a + c/2)/a^2(a + c)^2 \,, \qquad (4.51)$$

where the instantaneous reproducing head voltage e equals

$$vm \; \Sigma \; (d\varphi_r/dx).$$

Thus the total noise power, proportional to e_n^2, is proportional to the track width, w, and the square of the tape velocity, v. Also, for a fixed tape magnetization, the quantity, Dm^2, is proportional to the particle volume and lower noise power would result from a reduction of particle size.

The spectral distribution of the above calculated total noise voltage may be determined by expanding the head flux response function as a

Fourier integral, e.g. HOWLING [1956]. It is found that the noise voltage appears to be due to a noise magnetization of constant intensity over the frequency spectrum if the tape cross sectional area is considered to be infinitely small. In this case only the white noise of the tape is modified by the frequency response of the head to sinusoidal signals, to give the reproducing head voltage. For practical conditions, with finite tape cross-section, the mean square noise must be averaged over the tape cross section, resulting in a similar but not identical spectral distribution to that of the signal (BROWN [1956a]). Thus the earlier assumption that the tape noise spectrum may be determined from the head response to sinusoidal signals and the noise voltage spectrum is only approximately correct. The analysis of background noise gives reasonable agreement with the experimental amplitudes measured on commercial tapes. Thus the attribution of background noise to the particulate nature of the tape appears to be basically correct. However, the above analysis cannot predict the observed increase of noise with tape magnetization level, or modulation noise.

§ 6.2. *Modulation Noise*

Modulation noise, sometimes called "the noise behind the signal", is that part of tape noise whose amplitude is a function of the magnetization level. It is significantly larger than the background noise. In audio recording the modulation noise tends to be masked by the signal but it is particularly noticeable when the recorded signal consists of a single frequency. Two sources of modulation noise have already been mentioned, being caused by random fluctuations in the longitudinal and perpendicular motion of the tape, and leading to frequency and amplitude modulation noises, respectively. Experimentally, the major source of modulation noise appears to be due to spurious amplitude modulation of the recorded signal by variations in the physical and magnetic properties of the tape (PRICE [1958]). Non-uniformities in the surface of the tape predominantly affect the short wavelength components of the recorded signal and are responsible for the frequency dependent component of the modulation noise. The frequency independent component is caused by physical and magnetic non-uniformities in the coating and is operative at all signal wavelengths.

The experimentally observed dependence of noise on tape magnetization shows greatest increase, compared to a demagnetized tape, in the long wavelength noise components. One contributor to this noise is the unevenness of the tape backing material which leads to varying coating thicknesses. Due to this there is a non-uniform recorded flux for recording signals less than those required to produce tape saturation (SMALLER [1959]). Particle agglomeration in the tape is another cause of modulation noise.

The ideal model used for the analysis of background noise assumes no particle agglomeration and predicts a lower noise level for a magnetized tape than for a demagnetized tape, in contradiction to experiment. Agglomerations of particles can be looked upon as being equivalent to changes in particle size if the increased interaction between the particles results in correlation of the particle magnetization directions. Plausible models for correlation between pairs of particles are described in Chapter 2, § 5, showing that the relative magnetization directions depend on the directions of the external demagnetizing fields of one particle on another. Thus, effectively, the correlation of particle magnetization directions is increased in agglomerations due to the large interaction fields. Hence, when the magnetization of those agglomerates, which formed antiparallel or other internally closed magnetization paths, is increased, the particle magnetization directions are forced into parallel alignment leading to an increase of their external fields (STEIN [1962]). Finally, the inverse of particle agglomeration, that is voids or holes in the coating, can also contribute a magnetization dependent noise. A random distribution of identical holes in a continuous magnetic medium is equivalent to a similar set of magnetic particles magnetized in the opposite direction to the medium. This may be analysed in a similar way to the model for ideal particles described above to explain background noise. In the case of holes, however, the equivalent opposite magnetization increases with the magnetization of the surrounding medium and hence their noise contribution also increases.

It may be concluded that the physical bases for the modulation noise differ from the particulate nature of the tape particles which account for background noise. Both particle agglomerations and holes in the coating can account for the magnetization dependent modulation noise. Surface roughness of the tape may be looked upon as an increase of holes near the tape surface which contribute more noise at short wavelengths.

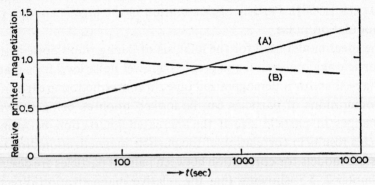

Fig. 4.21.   Time dependence of print-through magnetization.
(A) Time of exposure to printing field.
(B) Time after exposure to printing field.

§ 7. PRINT-THROUGH

In the category of reproduction of unwanted tape magnetization there is, in addition to the noise sources discussed in the preceding section, a special case of undesirable signal recording. This occurs while a recorded tape is stored in a wound reel, and is commonly known as print-through. The external tape field due to a recording is sometimes large enough in adjacent layers of the wound reel to produce a change in the magnetization direction of a few of the particles. The processes and conditions under which fields of smaller magnitude than the bulk coercivity can cause magnetization change are discussed in Chapter 5, § 2.4. The manifestation of the printed signal in reproduction will now be studied for a variety of printing conditions.

In Chapter 5 it is shown that, as the volume of the tape particles is reduced and the temperature increased, the relaxation times for magnetization change are reduced and the particles reach their terminal magnetization more rapidly. In this case the probability of magnetization by a small external field in a given time is increased. In a practical material a distribution of particle volumes and shapes will occur giving a corresponding distribution of relaxation times according to eq. 5.29 (Ch. 5, § 2.4). For a wide range of relaxation times the resulting

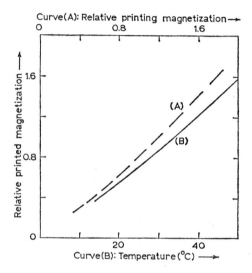

Fig. 4.22. Dependence of print-through magnetization on printing field and ambient temperature.

change of magnetization, after a change of applied field, is logarithmic (NÉEL [1949]), and of the form

$$I_r(t) = I_r(0) - AT \log t , \qquad (4.52)$$

where the remanent magnetization $I_r(0)$ at time $t = 0$ has decayed to $I_r(t)$ in time, t, in zero external field at temperature, T. The factor, A, is determined by the volume and shape distributions of the particles. The relaxation times are modified by the presence of a small external field, as described in Chapter 5, § 2.4, and an increase of magnetization occurs proportional to the ambient temperature and to the logarithm of time.

The above characteristics are exhibited by γ Fe_2O_3 powder tapes subjected to print-through fields for different time periods and at different temperatures. Fig. 4.21 shows the logarithmic increase of printed magnetization with exposure time and the corresponding decrease after exposure. Figure 4.22 illustrates the approximately linear dependence of printed magnetization on the amplitude of the printing field, plotted as magnetization in the printing tape, and on the ambient temperature. It is clear, then, that the unwanted printed

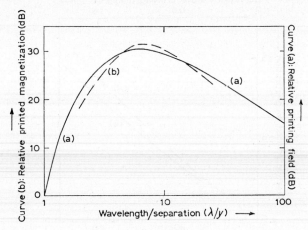

Fig. 4.23.   Wavelength dependence of printing field and printed magnetization.

signal can be minimized by avoiding temperature rises; similarly the printing field is reduced by the use of a thick plastic base, which increases the layer-to-layer separation in the wound reel. Unfortunately, the latter expedient decreases the length of tape which can be wound in a given volume whereas the present trend to higher information packing densities leads to thinner rather than thicker base materials. However, in some cases an alleviating factor, in high resolution systems, occurs due to the reduced printing field for short recorded wavelengths.

The wavelength dependence of the maximum longitudinal field at a distance, y, outside a sinusoidally recorded tape is given by

$$H_{x(\text{max})} = \frac{4\pi^2 c \, I''_{x(\text{max})}}{\lambda} \exp\left[-2\pi y/\lambda\right], \qquad (4.53)$$

where c and y are the coating and tape thicknesses, respectively, if the field is the layer-to-layer printing field. As the recorded wavelength decreases, an initial increase of the printing field, due to the increase of magnetic pole density, is offset by the exponential separation loss. The resulting wavelength dependence of the printing field is plotted in Fig. 4.23 – curve (a), on a decibel scale. It is found that the printed magnetization, curve (b), has a similar shape to the field curve, indicating the linear dependence of the flux on the applied field for

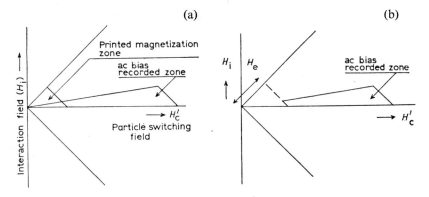

Fig. 4.24.   a. Preisach diagram of recorded and printed magnetization.
b. Diagram after ac erasure by field H_e.

low fields (DANIEL and AXON [1950], WESTMIJZE [1953c]). The maximum printed flux occurs when $\lambda = 2\pi y$ and thus moves to shorter wavelengths as the tape thickness is decreased. However, in some new recording systems, such as the wide-band high-resolution system described in Chapter 7, the maximum print-through occurs at the long wavelength end of the band even for thin tapes. Thus print-through effects are negligible in such cases, but may be considerable at high tape speeds.

Finally, the stability of the printed magnetization will be expected to be markedly inferior to the true signal magnetization since it exists in those particles that are inherently unstable. The unstable nature of the printed magnetization may be illustrated by the formal Preisach diagram representation (Chapter 2, § 5). As can be seen from Fig. 4.24a, the particles which become magnetized by the printing field are those with low intrinsic switching fields (H_c') and low interaction fields (H_i) included by the small triangle. On the other hand, if a similar field is recorded with ac bias, low interaction field particles through the whole range of intrinsic switching fields are magnetized. It is then clear that a small erasing field, of the same order of magnitude as the printing field can completely demagnetize the printed magnetization, while only a small part of the recorded signal is affected as shown in Fig. 4.24b. Whereas the Preisach diagram is not completely satisfactory physically (see Chapter 2, § 5), the idea of different particles, or parts of particles, being magnetized by small and large applied fields is advanced in a way which demonstrates the relative magnetization stabilities.

CHAPTER 5

MAGNETIC TAPE – THEORETICAL CONSIDERATIONS

The mechanisms of magnetic recording, reproduction and storage have been discussed in the previous chapters in terms of the practical magnetization characteristics obtained in current magnetic tapes. The formulation of the optimum tape properties for a given type of recording is then possible and is considered initially in this chapter. The physics of permanent magnetic materials is a separate subject, beyond the scope of this book, but developments in fine particle ferromagnetism are directly applicable to magnetic recording materials and these will be described in some detail. In fact, magnetic tape probably represents the largest commercial application for permanent magnet powders, and future tape developments rely largely on a better understanding of the factors controlling their magnetic properties. The degree to which practical magnetic materials can meet the requirements for an ideal tape material is studied in the next chapter.

§ 1. TAPE DESIGN CRITERIA

The essential function of the recording medium, used in conventional magnetic recording systems, is to produce, at the surface of the reproducing head, a magnetic field whose amplitude has the same time variation as that occurring during the recording process. The field can be looked upon as being due to a distribution of magnetic poles on the surface of the tape. Thus it is important that the tape surface be smooth and relatively soft to allow continuous physical contact between the tape and the recording or reproducing head, and thus ensure maximum interaction between the magnetic materials in the tape and head. Due to the finite thickness of the magnetic layer, magnetic poles inside the tape contribute a relatively smaller part of the tape surface field. This loss may be minimized by use of a high remanent magnetization material in a very thin layer. This, in turn, makes a high packing density of magnetic material desirable, a requirement which is in conflict with the required pliable nature of the coating for good head-to-tape contact.

Up to the present time, the general physical requirements for magnetic tape have been most adequately met with a magnetic layer consisting of a fine magnetic powder dispersed in a plastic binder and coated onto a thin plastic tape. All-metal tapes have certain restricted uses, where long wavelengths only are encountered, and where the gain in maximum tape surface field overrides the loss due to imperfect contact with the heads. A possible successful tape design may arise from the use of thin metallic layers on a plastic base material which could combine the advantages of physical flexibility and large surface field.

The general tape construction, consisting of a magnetic coating on a nonmagnetic carrier, also has the advantage that each layer may be separately optimized to fulfill its allotted function. The desirable features of good long-term physical stability, uniformity, flexibility and low density are all aimed at in the base material. Likewise, the coating is designed to yield optimum magnetic properties, along with a suitable physical surface which permits constant velocity of transportation past the head and intimate contact over the whole tape width. Both base and coating should be made as thin as possible in order to achieve a maximum volume density of information in a recording. This condition also makes for good tape flexibility. Provided equivalent physical properties can be maintained in thin base layers, the limiting factor will be the magnetic printing effect between adjacent layers in the wound tape.

For the magnetic coating, several conditions determine the overall optimum thickness apart from the attenuation due to separation of the magnetic material from the reproducing head. As shown in Chapter 2, the recording resolution diminishes with separation from the recording head since the recording field decrement in the direction of tape travel is lowered. Also, since the recording sensitivity is optimum for a critical bias field (Ch. 3, Fig. 3.14), a greater departure from simultaneous optimum conditions in all layers of the coating occurs for thicker coatings. All these effects depend on the signal wavelength, however, and when a range of frequencies is to be recorded, a compromise thickness is chosen so that the long wavelengths are limited by coating thickness and the short wavelengths by over-biassing. On the other hand, the improved efficiency of recording and reproducing with very thin magnetic layers has a limit of usefulness, since the long wavelength surface field reduces with thickness reduction whereas a corresponding overall noise reduction does not take place. This limit will be determined by the density of magnetization achieved in the magnetic material.

For the magnetic material in the tape it is desirable to achieve a maximum remanent intensity of magnetization in the direction of the recording field. In addition, since the field changes direction in the recording zone at the trailing edge of the recording head, it is also advantageous if the magnetization change is small due to fields in directions other than the preferred direction. The preferred direction is usually along the direction of relative motion between the tape and the head, for conventional "longitudinal" recording, even though the recording field direction rotates towards a perpendicular direction near the recording head. Very thin coatings for short wavelength recording may give larger remanent magnetizations if their preferred direction is tilted towards that of the recording field.

The coercivity of the ideal tape material is governed by conflicting requirements. Since it represents the force required to destroy the magnetization in the material, it is necessary that it be sufficiently high to minimize the loss of remanent magnetization due to the demagnetizing fields acting in a recorded tape. These fields can occur both from the self demagnetization of the magnetic particles, and from the variations of magnetization direction and magnitude of the recording. For magnetic oxide powder tapes the saturation tape magnetization is relatively low, due to the intrinsically low saturation magnetization of ferrites and the relatively low volume packing density in powder coatings. It will be shown that coercivities of the order of 300 Oe are sufficient to render demagnetization losses small. However, for very thin magnetic metal tapes the tape magnetization, and hence the optimum coercivity, may be increased by a factor of ten. A limit on the use of such material then exists in the power required to produce the high fields necessary to record and erase such a material. Furthermore, since the demagnetizing field depends on the recorded wavelength, the optimum coercivity may depend on the specific application, and it is to be expected that a relatively higher ratio of H_c/I_r'' will be necessary in very thin layers.

It has been implied, in the previous discussion, that a rectangular hysteresis loop is desirable in the tape, since then large remanent induction is achieved, along with minimum sensitivity to self-demagnetization. A rectangular hysteresis loop is obtained when the effective switching fields of the magnetic elements of the coating are alike. This property is, in itself, desirable since it improves the recording definition and minimizes the loss due to departure from optimum bias conditions in layers at different depths in the coating. In this latter respect it may be even more advantageous to grade the particle switching fields through the coating thickness to be similar to the field decrement so that all layers are optimally biassed. Finally, it is desirable that the magnetic material exhibit a linear anhysteretic remanent magnetization characteristic. However, this characteristic is thought to be controlled by random internal fields and is inherently non linear (Ch. 2, § 5); future improvements can be expected here.

The question of optimizing the magnetic field at the surface of the tape has so far been considered only from the point of view of obtaining maximum surface field for reproduction. In respect to the printing of signals between layers of a wound reel of recorded tape, a large field is undesirable. Thus, another requirement for the magnetic material of the tape is that it be insensitive to remagnetization by small dc fields which may be accompanied by other energies, such as temperature rise or external ac fields. It will be shown that magnetic stability of fine powders is a function of the particle size which thus becomes important in the design of a suitable magnetic material. Apart from determining the mode of magnetization change (superparamagnetism, domain vector rotation or domain wall motion) the particle size will also influence the noise level in reproduction for a tape subjected to bias but no signal. The effective particle size will also be influenced by the degree of dispersion of the magnetic elements, but, apart from this, lower noise levels can be expected from smaller particles. The optimum particle size is the smallest size compatible with magnetic stability.

The tape design considerations which have been discussed depend on how far the components of the tape can be made to meet the preferred characteristics. In the ensuing sections of this chapter the theoretical properties of magnetic powders and thin magnetic sheets are studied in detail; practical tapes are described in Ch. 6.

§ 2. MAGNETIZATION OF SMALL PARTICLES

The classical theory of Weiss postulates that ferromagnetic substances are divided into elementary domains in which the magnetization is uniform. The size of the domains and the direction of magnetization are the variables through which magnetization takes place at any given temperature. The magnetization within a domain of a given material is a function of temperature only and decreases with increasing temperature, becoming zero at a characteristic temperature for the material called the Curie point. Thus, the three processes by which magnetization changes can take place are:

(1) Change of magnitude of spontaneous magnetization with temperature.

(2) Change of direction of magnetization by rotation of spontaneous magnetization.

(3) Change of domain size by domain wall displacement.

Stable irreversible magnetization at relatively high applied fields is required for magnetic tape and this is obtained in so-called single domain particles where the primary mechanism of magnetization is by domain vector rotation. Recording can also take place by simultaneous application of a magnetic field and heat, although this technique is not widely used. The application of heat causes sufficient temperature rise to produce a change of spontaneous magnetization and this allows permanent magnetization due to the applied field to take place on cooling. Domain wall displacement is usually a low energy process, but when only a few walls are present in a sample, this can become a magnetization process in a magnetically hard material. Thus all three magnetization processes are pertinent to magnetic recording. The first and third processes are of incidental importance and will be briefly discussed before a detailed analysis is made of domain vector rotation processes.

§ 2.1. *Magnetization Mechanisms in Small Particles.*

Spontaneous Magnetization

The single domain particles used for magnetic tape are conventionally ferrimagnetic oxides, although ferromagnetic metals and alloy powders also exhibit suitable permanent magnet properties. In the ferromagnetic powders (e.g. iron), all the atoms are identical and have the same magnetic moment. The energy of coupling between adjacent moments is a minimum when the moments are oriented in the same direction and so the spontaneous magnetization is given by the summation of the elementary moments in unit volume. A more complicated situation occurs in the case of the ferrimagnetic oxides. For many ferrites exhibiting ferrimagnetism the crystal structure is inverse spinel, in which the trivalent metal ions may occupy either of two sites in the oxygen ion lattice. Site A refers to a tetrahedral position surrounded by four oxygen ions and site B to an octahedral position surrounded by six oxygen ions. For each molecule, one site A and two sites B exist. The trivalent iron ions, always present in ferrites, occupy one of each site in inverse spinel and such structures are magnetic. The divalent metal ion occupies the other B site. The strongest coupling energy is between the A and B sites *via* the oxygen ions and is a minimum when the moments are aligned in opposite directions. Thus, since the trivalent ions exist in both sites, their resultant magnetic moment is zero and the spontaneous magnetization is the resultant moment per unit volume of the divalent ions. Thus, for ferrous ferrite (magnetite), a moment of 4 Bohr magnetons per molecule is obtained, the value for cobalt ferrite being 3 Bohr magnetons; these are the values of the spin magnetic moments of Fe and Co respectively. Likewise, nickel and copper ferrites have magnetic moments of 2 and 1 Bohr magnetons, respectively. Experimental values are found in fairly good agreement with the simple model outlined. Another class of permanent magnet ferrites has a hexagonal structure. In these materials the oxygen ions form a hexagonal lattice with some ions substituted by Ba, Sr, Pb or Ca ions; they may also contain interstitial divalent ions. Possible tape materials having inverse spinel and hexagonal structures are described later in this chapter.

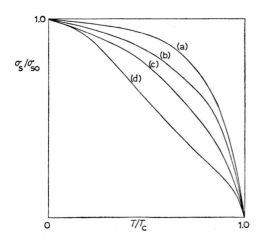

Fig. 5.1. Normalized magnetization *vs.* temperature characteristic for various magnetic metals.
(a) Ferromagnetic metals.
(b) Fe_3O_4.
(c) $CoFe_2O_4$.
(d) $BaFe_{12}O_{19}$.

The temperature dependences of spontaneous magnetization for ferromagnetic metals and alloys have the general shape shown in Fig. 5.1 – curve (a), which is a monotonic decrease to zero at the Curie point. All simple ferrites with spinel structure exhibit the same type of temperature dependence, with a small variation of specific saturation magnetization (σ_s) at low temperatures becoming very steep near the Curie point. The normalized curves for magnetite and cobalt ferrite are shown in Fig. 5.1 (PAUTHENET [1950]). In mixed ferrites, however, the bulk magnetization may be due to the algebraic addition of different magnetic moments arising from the magnetic ions located in different sites. Furthermore, each network or sublattice of ions, having a coherent direction of spontaneous magnetization, can have different rates of fall of spontaneous magnetization with temperature. This results in an anomalous characteristic for the bulk material which can have maxima and minima. Most hexagonal ferrites exhibit a somewhat more linear temperature dependence of magnetization than spinels or metallic ferromagnetics. The normalized temperature dependence of spontaneous magnetization for barium ferrite is also shown in Fig. 5.1 – curve (d).

As described in Chapter 3, § 3.2, irreversible magnetization at elevated temperatures requires relatively small applied fields as the particle switching fields reduce to zero near the Curie point. However, for some materials, the switching fields can fall very rapidly with a temperature rise from room temperature, yielding the possibility of high sensitivity to recording fields at temperatures below the Curie point. Such is the case for cobalt ferrite, (Fig. 3.20), whereas for barium ferrite the switching field is practically invariant with temperature. This widely different behaviour is due to the different temperature dependences of crystal anisotropy (K_1) for the two materials. The switching fields are proportional to K_1/I_s; however, for spinels like cobalt ferrite exhibiting cubic anistropy, K_1 decreases very rapidly with temperature rise, compared to the change of I_s. For uniaxial crystal anisotropy, on the other hand, the variation is much less, and for barium ferrite the crystal anisotropy is nearly proportional to I_s.

Since the spontaneous magnetization is due to strong alignment forces between neighbouring metal ions, it is evident that, as the volume of the material is reduced, a condition is reached where the alignment forces are smaller than the randomizing forces due to the ambient temperature. In such cases thermal fluctuations cause the magnetic moment to change spontaneously from one direction to another. This is a time effect which may be described in terms of a relaxation time τ_0 where

$$I_r(t) = I_r(0) \exp\left(-t/\tau_0\right), \tag{5.1}$$

considering an assemblage of identical particles.

When τ_0 becomes short compared to the time of the experiment, ferromagnetism is effectively lost and the material is said to be super-paramagnetic, since it exhibits paramagnetic behaviour and the particles have large magnetic moment. The magnetization in super-paramagnetic particles is achieved reversibly and can have a large initial susceptibility. A typical magnetization curve (BEAN [1955]), for a random assembly of such particles is shown in Fig. 5.2a. The experimental magnetization curve for small particles of magnetite made by Lefort's method (ELMORE [1938]), is shown in Fig. 5.2b, indicating

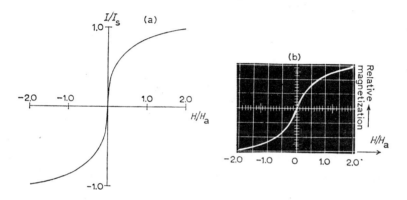

Fig. 5.2.   Magnetization characteristics for superparamagnetic particles:
a. Theoretical magnetization curve for superparamagnetic particles (BEAN [1955]). H_a — anisotropy field = $2 K/I_s$; K — anisotropy constant.
b. Experimental magnetization curve. Colloidal magnetic particle diameter $\approx$ 100 Å. $H_a \approx$ 500 Oe.

superparamagnetic behaviour as described. This instability of magnetization in very small particles has not found application for recording purposes, although it is probably a contributor to accidental recording or "print-through" (§ 2.4.).

Dependence on Particle Size

When the volume of a particle is large enough, the elementary magnetic moment alignment forces override the thermal fluctuations and relatively large areas of the particle are coherently magnetized. For ferric oxide at room temperature, this condition is obtained for particles whose volume is greater than about 10^{-17}cc. Magnetization may then, in general, take place by movement of the boundaries (domain walls) between the coherently magnetized areas (domains), or by rotation of the magnetization directions of the elementary magnetic moments (the uncompensated electron spins). Domain wall movement in bulk material is usually a lower energy process than magnetization rotation and is the principal mode of magnetization in large particles. However, as will be described later, a critical size occurs below which domain walls are energetically unfavourable and such particles are

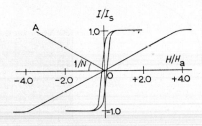

Fig. 5.3a.   Effect of demagnetizing field on hysteresis loop of multidomain material.

referred to as single domain particles. These particles are of greatest interest for the permanent magnet properties required for magnetic recording. However, in the region of the critical size corresponding to the onset of multidomain behaviour, large energies are required for magnetization even though a few domain walls may exist. Such multidomain particles are certain to exist in any practical powder with a distribution of particle sizes. Powders for recording purposes preferably should have a large single domain range: that is to say, the critical sizes for single domain behaviour, with respect to superparamagnetic or multidomain behaviour, should be well separated.

When particles are large enough for domain wall formation to become energetically favourable, a relatively low coercive force is obtained due to the lower energy required for wall displacement than for domain vector rotation. As the particle size is reduced, it is observed experimentally that the coercive force increases in approximate inverse proportionality to the particle diameter, d, tending towards a limiting value for single domain particles. The coercive force is determined by the field required to nucleate domains of reverse magnetization after saturation, and also by the field required to propagate the walls of the reverse domains. Multidomain particles are difficult to saturate since the applied field has to override the demagnetizing field of the particle. It has been shown (FOWLER et al., [1961]), that, for pure iron single crystals in the form of multidomain whiskers, an applied field of 6000

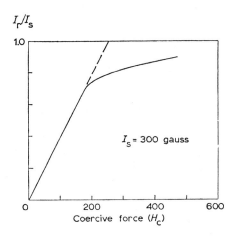

Fig. 5.3b.   Dependence of remanent magnetization on coercivity for cubic shaped particles of iron oxide (KRONES [1960]).

Oe is insufficient to override the demagnetizing fields at the corners of the whiskers where subsidiary domains persist.

The demagnetizing field coefficient, N ($= 4\pi/3$ for a sphere) can have a large influence on the remanent magnetization of particles, especially for large particles with low coercive forces. Fig. 5.3a shows the magnetization curves for a multidomain bulk material and also for the material under the influence of a demagnetizing field. The remanent magnetization I_r is given by the intersection of the descending branch of the hysteresis cycle with a line of slope $1/N$ passing through the origin. When N becomes larger than H_c/I_s, then I_r approaches H_c/N and

$$I_r/I_s = H_c/NI_s . \tag{5.2}$$

This relationship is only true for low values of I_r/I_s where direct proportionality between I/I_s and H/H_a exists. For large values of H_c, the demagnetization line OA intersects the curved section of the hysteresis cycle and the remanent magnetization increases less rapidly with increase of H_c. The dependence of I_r/I_s on H_c is shown in Fig. 5.3*b* for cubic shaped particles of iron oxide (KRONES [1960]). It can be seen for this material ($I_s = 300$) that the demagnetization field

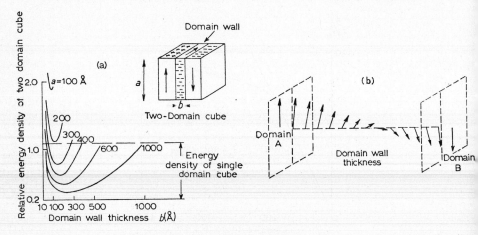

Fig. 5.4.   Two-domain particle model:
a. Energy of two-domain cube (AMAR [1958]).
b. Directions of electron spins in wall between two domains magnetized oppositely.

produces only a small loss of apparent remanent magnetization, providing the coercivity exceeds 300 Oe.

As the size of a magnetic particle is reduced, the number of domain walls is also reduced and the finite boundaries of the particle begin to change the conditions leading to minimum total energy of the particle. In a model consisting of a two-domain cube, it has been shown (AMAR [1958]) that the width of the domain wall, for minimum total energy, is less than in the bulk material (see Fig. 5.4*a*). As shown in the family of curves of total magnetic energy of such a cube of iron, the energy minima for particles between 300 Å and 1000 Å size correspond to domain wall widths in the range 145 Å to 225 Å, whereas a width of 1000 Å is obtained in bulk iron. Furthermore, due to the reduction in wall width, the angle between adjacent electron spins in the wall will be increased (Fig. 5.4*b*), and the exchange energy between them increased. This leads to an increase of the wall energy per unit area compared to bulk material. Thus, in very small particles, the minimum energy state will be achieved for structures with few domain walls. Eventually, as shown in Fig. 5.4*a*, the minimum energy condition is

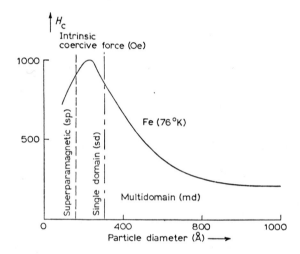

Fig. 5.5.   Coercive force for iron powder as a function of particle diameter (MEIKLEJOHN [1953]).

obtained with no domain walls, since the total energy of a uniformly magnetized particle (dotted line, Fig. 5.4*a*) is less than that of the two-domain particle. As can be seen from the figure, this condition is obtained in iron for a cubic particle whose side is less that 200 Å and single domain behaviour is energetically favourable.

Somewhat more complicated zero-field domain structures might be expected in non-cubic shaped particles and in particles with cubic rather than uniaxial anisotropy. However, the magnetization characteristics are not materially altered from the simple two-domain particle considered above, since they are largely controlled by the increase of magnetostatic energy on increasing the unidirectional magnetization. On the other hand, the transitional magnetization structures in the region of the critical size for single domain behaviour are not easy to predict. Much depends on the thickness of the domain wall compared to the size of the particle. For some materials this ratio is small and the simple picture of Fig. 5.4 is applicable. However, for medium aniso-tropy materials, like iron and nickel, the domain wall covers the whole particle and the concept of coherent magnetization has to be abandoned. Indeed, incoherently magnetized structures are also possible during magnetization change in "single domain" particles considered in the next section.

This discussion of multidomain (md), single domain (sd), and superparamagnetic (sp) structures in fine powders may be summarized by considering the experimental dependence of coercive force on particle size as shown in Fig. 5.5 (MEIKLEJOHN [1953]).

The three ranges of magnetization structure are indicated. Whereas the boundary between (md) and (sd) structures is theoretically quite sharp, no such distinction can be made between (sd) and (sp). It is more correct to think of the conventional domain structure as a special condition of a general structure of incoherent magnetization.

As the particle size is reduced, this special condition is modified and the energy density of the particle increases, leading to an increase of coercivity. These modifications result in changes of domain wall thickness and energy, and a continuous directional variation of the magnetization vector occurs throughout the particle. However, another special condition may occur as the particle size is reduced before thermal fluctuations destroy coherent magnetization. In this condition minimum total energy corresponds to minimum exchange energy and the magnetization vector is unidirectional throughout the particle, at least in zero external field. Maximum energy density occurs in this range (sd). The process of magnetization change is governed by the anisotropies of shape, stress, and crystal energy, and will be discussed in more detail in the next section.

§ 2.2. *Magnetization Models for Single Domain Particles.*
   Uniaxial Anisotropy

Although magnetization change in small particle multidomain

structures requires large applied fields, this process is undesirable in a material for magnetic recording since it is essentially linear. A non-linear characteristic is required for recording, leading to high sensitivity to recording signals in the presence of a suitable bias and low sensitivity to magnetization change after recording. An irreversible magnetization process controlled by some critical energy is therefore required.

Fortunately, this linear process in multidomain particles, whose magnetization is controlled by the magnetostatic energy, can be avoided by using very small particles. In this case, exchange energy minimization is a dominant factor and only small changes of magnetization vector direction are allowed in zero applied field. Thus the large magnetostatic energy, supplied by the field for multidomain particles, is permanently stored in small particles. The magnetization characteristic is then controlled by the internal anisotropies of the particles which oppose the rotation of the magnetization vector towards the field direction. It is desirable that the controlling anisotropy be so oriented that, after saturating the material, a minimum magnetization change occurs on removing the field. This applies to both irreversible and reversible changes. However, on increasing the applied field in the opposite direction to the magnetization, minimum magnetization change should occur until a critical field is reached. At this point, maximum irreversible magnetization should take place in a small field range. The average of the applied fields over which these irreversible changes occur is called the coercive force.

The mechanisms for magnetization vector rotation will now be described for simple models of magnetic particles. Since the particle boundaries have an important effect on the magnetization characteristic, simple shapes will be considered viz. cylinders, spheres and ellipsoids. In this section the simplest form of anisotropy, uniaxial anisotropy, will be used.

Coherent Rotation

The simplest mechanism for rotation of the magnetization vector is that in which all the electron spins in a particle are parallel at all times. This is an idealized case, but is an extremely useful and successful model. Detailed calculations have been made with this model and with extensions to the model (STONER and WOHLFARTH [1948], WOHLFARTH [1956 ,1959a]). The internal particle energy of a magnetization vector I_s depends only on the orientation of I_s with respect to particular axes in the particle. For a simple magnetically uniaxial particle the internal energy $E_i(\theta)$ is given by

$$E_i(\theta) = Kv \sin^2 \theta ,\qquad (5.3)$$

where K is a constant, v is the volume of the particle and θ the angle between I_s and the axis. The constant, K, is the anisotropy constant for the particle and can be due to three different kinds of energy.

(1)   Magnetocrystalline energy

If the particle is a single crystal with uniaxial anisotropy, eq. (5.3) applies (neglecting a constant and high order terms) where θ is the angle between the magnetization and the single axis of preferred magnetization. In this case $K = K_1$ in the expression for the magnetocrystalline energy density E_c', where

$$E_c' = K_0 + K_1 \sin^2 \theta + K_2 \sin^4 \theta \ldots. \qquad (5.4)$$

(2)   Anisotropic internal field energy

If the particle is ellipsoidal in shape, a uniform internal field is produced with components

$$H_i = - N_i I_s . \qquad (5.5)$$

Here, N_i depends only on the ratio of the semi-axes a_i. For a prolate spheroidal particle with principle semi-axes a, b, b, where $a > b$, the demagnetizing factors corresponding to magnetization along the polar and equatorial axes N_a, N_b are related by

$$N_a + 2N_b = 4\pi . \qquad (5.6)$$

Since $N_a < N_b$, the polar axis is the direction of easy magnetization and the energy associated with the demagnetizing field H_i is given, apart from a constant, by

$$\begin{aligned} E_i(\theta) &= \tfrac{1}{2}vN_iI_s^2 \\ &= \tfrac{1}{2}vI_s^2 \left[N_b - N_a\right] \sin^2 \theta. \end{aligned} \qquad (5.7)$$

Again, θ is the angle between the easy axis and the magnetization, and eq. (5.7) is similar to eq. (5.3) with $K = \frac{1}{2}I_s^2[N_b - N_a]$.

(3) Magnetoelastic Energy

The interaction between the magnetization of a particle and the mechanical stress acting on it is determined by the change in magneto-crystalline anisotropy associated with the stress. This relationship between crystal and stress anisotropy leads to similar formulae for the energies associated with them. For a uniform mechanical stress, T, and an isotropic saturation magnetostriction constant (λ_s), the magnetoelastic energy density, E_m', is given by

$$E_m' = \tfrac{3}{2}\lambda_s T \sin^2\theta \,, \tag{5.8}$$

thus K in eq. (5.3) is given by $K = \tfrac{3}{2}\lambda_s T$.

When an external magnetic field H is applied to a particle with uniform magnetization I_s, external magnetic energy is associated with the interaction between them, given by

$$E_e(\psi) = - v H I_s \cos \psi \,, \tag{5.9}$$

where ψ is the angle between H and I_s. The equilibrium magnetization state for a particle in an applied field H will be that direction of magnetization giving minimum total energy E_t, where

$$E_t = E_e + E_i \,. \tag{5.10}$$

Calculations of the dependence of the magnetization, in the applied field direction, $I(\theta + \psi)$, on the magnitude of the field have been made (STONER and WOHLFARTH [1948]) and curves for $\theta + \psi = 0°$, $20°$, and $40°$ are shown in Fig. 2.9. Thus the desirable rectangular hysteresis loop is predicted when the applied field is close to the particle axis. Whatever form of anisotropy is used in practice, it will be necessary, then, to orientate the easy magnetization axis in the predominant field direction of the recording head.

Further comparison with practical powders may be made by superposing the magnetization curves for particles with various orientations to simulate an unoriented powder sample. In general, for a particle directed at an angle to the applied field, a critical field, H_c', occurs at which the magnetization suddenly rotates irreversibly towards the applied field direction. Reversible magnetization then occurs slowly up to saturation. For particles with θ small, H_c' is large and little reversible magnetization occurs, whereas for θ large, H_c' is small and the magnetization change is mostly reversible. A diagramatic representation of the

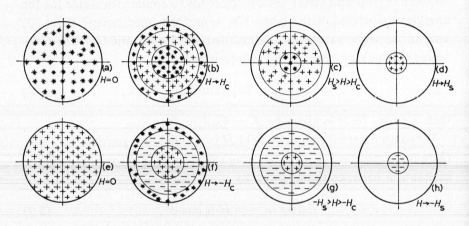

Fig. 5.6. Polar diagram of magnetization process in single domain particles of random orientation. (Applied field perpendicular to sheet.)

acquisition of magnetization by an unoriented array of acicular particles is given, in Fig. 5.6, by means of polar plots in which the pole lies along the applied field direction. From the demagnetized state (a), represented by stars, a positive field produces first a rotation of magnetization vectors (denoted by plus signs) towards the solid angle of 45°(b). Further increase of the field enhances this tendency by causing irreversible jumps of magnetization vectors from around $3\pi/4$ to $\pi/4$ and rotation away from $\pi/2$ (clear area (c)). As H approaches H_s the critical fields of particles whose vectors are oriented between $3\pi/4$ and π are reached, and all vectors are concentrated in a small cone around the applied field direction (d). Removal of the field produces unidirectional magnetization over the hemisphere corresponding to the remanence point (e). Increasing the field in the opposite direction then produces magnetization changes similar to (*b*) – (*d*) except that the centre core is positively magnetized rather than demagnetized. Quantitatively, the magnetization for any applied field at angle $\phi = \theta + \psi$ to the axis of the particle is given by

$$I''(H) = \int_0^{\pi/2} I''(H, \phi) \sin \phi \, d\phi . \qquad (5.11)$$

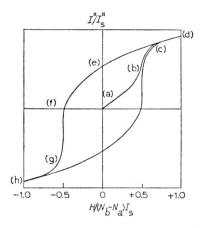

Fig. 5.7. Theoretical magnetization curve. Random distribution of uniaxial anisotropy. Coherent rotation model (STONER and WOHLFARTH [1948]).

This relationship is plotted in Fig. 5.7 using reduced coordinates I''/I_s'' and $H/(N_b - N_a)I_s$ and the points corresponding to Fig. 5.6 are marked. For this model, the coercive force and remanent magnetization are given by

$$H_c = 0.96 \, K/I_s ,$$

and

$$I_r'' = 0.5 \, I_s'' . \qquad (5.12)$$

In general, the coercive forces predicted by the simple theory are not attained in practice. There are several modifications which, in part, account for this discrepancy. As discussed in Ch. 2, § 5 particle interactions often cause a reduction of coercive force. The particles themselves may depart from the idealized picture of uniformly magnetized regions with uniaxial anisotropy. They may, for instance, contain a mixture of anisotropic energies, uniaxial and non-uniaxial; these modifications will be studied in a subsequent section. In addition, variations in individual switching fields of particles can cause the bulk coercive

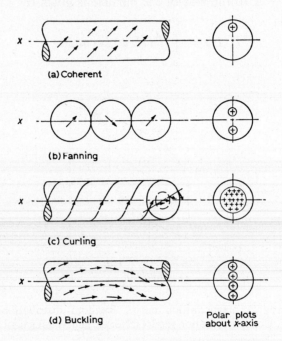

(a) Coherent

(b) Fanning

(c) Curling

(d) Buckling

Polar plots about x-axis

Fig. 5.8.   Schematic representation of modes of magnetization rotation.

force to be lower than that due to identical particles having a coercive force equivalent to the mean particle coercive force. Practical powders will have a distribution of particle sizes, and multidomain and intermediate magnetic structures should be accounted for. In particular, particles may be coherently magnetized in their stable magnetization states but may change their magnetization by incoherent rotation. This important mechanism for magnetization change will be considered in more detail.

Incoherent Rotation

As will be illustrated in the section on materials (Ch. 6, § 1) there is still a considerable discrepancy between practical and theoretical coercive forces for single particles with predominant shape anisotropy. The reason for this is that the coherent mechanism of rotation of the magnetization vectors considered (Fig. 5.8a) is not, in all cases, the lowest energy mechanism. Other, possibly lower energy, mechanisms have been proposed and have been given the names fanning, curling and buckling (see for instance JACOBS and BEAN [1955], BROWN [1957, 1959a], FREI, SHTRIKMAN and TREVES [1957]). The mechanisms are illustrated in Fig. 5.8b – d in which it can be seen that the fanning mode occurs in particles which have the form of a chain of spheres. Coherent rotation takes place in each sphere but in the opposite sense for adjacent spheres. Curling may be understood by imagining a parallel bunch of wires to be twisted. The direction of the wires then gives the direction of magnetization. Finally, for buckling, a sinusoidal variation of the magnetization vector takes place along the original direction of magnetization in a plane containing this direction. Such mechanisms represent solutions to a more general mathematical description of the equilibrium states of a magnetic material which do not invoke the idea of domains as fundamental. Rather, a general relation is used which equates the torque acting on the magnetization due to exchange and anisotropy forces, plus external and internal fields, to zero. Solving this equation and taking into account the boundary conditions allows the description of the magnetic structure of the material for any applied field. For so called "single domain" particles (i.e. coherently magnetized in zero applied field) magnetization change occurs by coherent rotation or by curling, depending on which is the lower energy mechanism. Since coherent rotation implies use of the field energy to overcome anisotropy and internal field forces, the characteristic field for onset of this mode is independent of the particle size. On the other hand, the curling mechanism involves only exchange and anisotropy forces since the internal field energy is unchanged during the process. The field required to overcome exchange forces depends on the particle size, however, and the field required to initiate the curling mode of magnetization change decreases with increasing particle size. The resulting theoretical critical fields, as a function of particle size, are shown in Fig. 5.9 for coherent and curling modes of magnetization change in an infinite cylinder. The operative mechanism will be the one having the lowest critical field. The curves for buckling and fanning modes of magnetization change are also plotted for comparison. The buckling mode is considered to be somewhat similar to coherent rotation but with a sinusoidal variation along the axis of the cylinder.

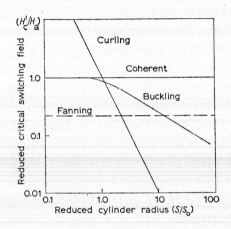

Fig. 5.9. Variation of coercive force with radius of infinite cylinder for different magnetization rotation modes (FREI et al. [1957], JACOBS et al. [1955]).
$H_a = (N_b - N_a) I_s = 2\pi I_s$; $S_0 = A^{\frac{1}{2}}/I_s$; $A =$ exchange energy constant.

Fanning is a special case for a chain of spheres, rather than a cylinder, which seems to be applicable to electrodeposited iron particles (LUBORSKY [1961]). It has also been shown (AHARONI [1959]), that, for finite particles, curling and coherent rotation are the preferred mechanisms for magnetization reversals for large and small particles, respectively.

The hysteresis loop for both mechanisms, for an applied field parallel to an infinite cylinder, consists of two stable states, coherently magnetized along the axis, separated by sudden jumps from one state to the other at the critical fields. These critical fields are different for the two mechanisms except for one critical radius of the cylinder. Thus the major feature of the incoherent rotation mechanism is that it predicts lower coercive forces, under certain conditions, than the coherent mode. This occurs for larger particles where the long range energies dominate. Magnetostatic energy is the only long range effect and it is minimized in an incoherently magnetized structure.

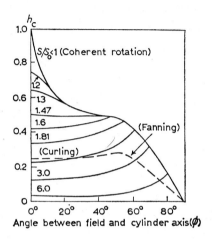

Fig. 5.10.   Angular variation of coercive force for infinite cylinder.
h_c – reduced coercive force = $H_c/2\pi I_s$ (SHTRIKMAN and TREVES [1959]).
See Fig. 5.9.

Coercive Force

The magnetization process has been described for single domain particles, with uniaxial anisotropy, in terms of specific mechanisms for the rotation of the magnetization from one direction to another. The two mechanisms called curling and coherent rotation appear to be basic, with buckling and fanning occurring for certain particle geometries. Since the angular variation of coercive force for a sample with aligned anisotropy is quite different for the two basic mechanisms, this characteristic is often used to indicate which mechanism predominates. The coercive force is also dependent on the interactions between particles and these will be discussed for the two mechanisms.

As far as is known, no stable magnetization states occur during the irreversible changes in a cycle. The critical field is then equal to the coercive force if the magnetization is positive, that is, in the opposite direction to the applied field, when this point is reached. The dependence of the coercive force on the angle (ϕ) between the applied field and the cylinder axis is shown in Fig. 5.10 (SHTRIKMAN and TREVES [1959]), for various values of the reduced cylinder radius S/S_0. The curve for fanning in a chain of spheres is also given. When the radius approaches zero, coherent rotation takes place at all angles. For larger radii, curling occurs for small angles and coherent rotation for large angles. A maximum in the angular variation of coercivity in practical materials is usually taken as an indication of incoherent magnetization processes. For partially aligned uniaxial anisotropy particles (JACOBS and LUBORSKY [1957], WOHLFARTH [1959b]), finite values of coercivity are predicted perpendicular to the average alignment direction. However, the zero value given by the models of Fig. 5.10 has been approached in highly oriented stainless steel wires (RASSMANN and HENKEL [1961]). In general support of the effects of partial alignment of uniaxially anisotropic particles, it may be noted that a large change in coercivity of the type predicted for coherent rotation is only observed in materials with a high degree of easy axis alignment. Such is the case for the permanent magnet alloy 'Ticonal XX'. However, the difference is due to an increase of coercivity in the alignment direction, rather than a decrease in the perpendicular direction, when compared to other alloys (e.g. Alnico) in the same family. This important result indicates that the high coercive force predicted by coherent rotation can be obtained in practice. As will be seen, however, most powder materials suitable for magnetic tapes and permanent magnets exhibit the lower coercive forces of the incoherent mechanism. Even for perfect single crystals, the theoretical value for an applied field along the uniaxial axis ($H_c = 2K/I_s$) is seldom observed, perhaps due to localized large demagnetizing fields at the corners of the particles.

From this discussion, it can be appreciated that the coercivity of practical powders depends on many factors. Two basic mechanisms of magnetization change are possible. Imperfections of alignment and in the particles themselves account for a general lowering of coercivities compared to the theoretical values. Interactions between particles also contribute to coercivity reductions (Ch. 2, § 5). For magnetic tapes, high coercivities *per se* are not required. It is important, however, that the spread of individual particle switching fields be minimized and so the magnetization mechanisms have to be understood.

Remanent Magnetization

For uniaxial anisotropy particles with random orientation, the remanent magnetization is equally distributed over a hemisphere giving a component in the saturation direction equal to half saturation (eq. (5.12)). Since it is important in magnetic tape to realize a large value of remanent magnetization, it is usually necessary to orientate the particles in the tape coating. In practice, sufficient orientation is normally achieved for the reduced remanence I_r''/I_s'' to reach values about 0.8.

The remanent magnetization for a random array of cylindrical particles is given by eq. (5.11) for an applied field (H) equivalent to the critical field for irreversible magnetization at an angle ϕ between H and the particle axis. Then,

$$I_r''(H) = \int_0^\phi I_s'' \cos \phi \sin \phi \, d\phi \,,$$

giving

$$I_r''(H)/I_s'' = \tfrac{1}{2} \sin^2\phi \,,$$

$$(5.13)$$

if the critical angle between the field and the cylinder axis increases monotonically with applied field. This is the case if the reduced radius of the cylinder S/S_0 is greater than 1.47, as shown in Fig. 5.10. The remanent magnetization curves may then be calculated (AHARONI [1959]), taking into account that coherent magnetization processes take place for large angles ($\phi > \phi_c$) between the field and the cylinder axes, and that curling takes place for $\phi < \phi_c$, where ϕ_c is the critical angle corresponding to equal critical fields for curling and coherent rotation. The theoretical remanent magnetization curves for $S/S_0 = 1.47$, 2.2 and 6.0 are shown in Fig. 5.11. Since the critical field for onset of irreversible magnetization due to curling decreases with the angle between the field and the cylinder axes, initial irreversible magnetization is due to particles aligned near the field direction. Thus, alignment of the particles towards the field produces a similarly shaped initial remanent magnetization curve with higher sensitivity, reaching remanence in a lower applied field. For very small radii ($S/S_0 < 1.47$), the magnetization for a randomly oriented sample is due to both curling and coherent rotation; the former gives way to the latter as S/S_0 decreases. The limiting remanent magnetization curve due to coherent rotation alone (for $S/S_0 \rightarrow 0$) is shown as a dotted curve in Fig. 5.11. If

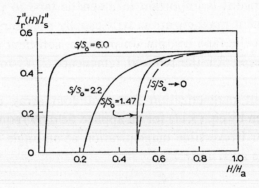

Fig. 5.11.   Theoretical remanent magnetization curves for random array of cylinders (AHARONI [1959]).
$H_a = 2\pi I_s$. S/S_0 = reduced cylinder radius. *See* Fig. 5.9.

allowance is made for variations of anisotropy, the curves are slightly less steep and have a toe giving gradual approach to zero remanent magnetization.

The angular variation of remanent magnetization is another relationship of interest in magnetic recording. Apart from its effect in recording, where the applied field varies in direction during the recording process (Ch. 2, § 6), there are important applications of magnetic recording where the direction of recording changes from one position to another. For instance, in recording on spinning magnetic disks (PEARSON [1961]), high remanence is required in all directions in the plane of the disk. It has been shown (DANIEL [1960]), that the degree of particle orientation obtained in magnetic tapes can be empirically described by a fourth power cosine function of the angle (ϕ) between the applied field and the mean orientation direction. In this case, the ratio of the remanent magnetization parallel and perpendicular to the preferred orientation direction is 8/3. As will be discussed later, a high ratio I_r''/I_s'' coupled with independence of I_r'' from ϕ can be achieved in powders with cubic anisotropy.

The remanent magnetization, $I_r''(H)$, considered thus far is that which has been acquired statically, that is by application and removal of a dc field H. In magnetic recording the remanent magnetization $I_{ar}''(H_{dc})$ is acquired anhysteretically (Ch. 2, § 4) when using ac bias. Static magnetization occurs in zero bias recording and yet another method of acquisition is used for dc bias recording. Here saturation by a dc field is succeeded by the application and removal of a dc field in the reverse direction and a remanent magnetization $I_d''(H_{dc})$ remains. A further method of remanent magnetization acquisition $I_d''(H_{ac})$ which has found occasional application in recording is by ac demagnetization of a previously acquired remanence from a dc saturating field. Relations between these remanent magnetizations have been derived for the simple model of non-interacting uniaxial particles (WOHLFARTH [1958, 1959c]),

$$I_d''(H_{dc}) = I_r''(\infty) - 2I_r''(H_{dc}) , \qquad (5.14)$$

and

$$I_d''(H_{ac}) = I_r''(\infty) - I_r''(H_{dc}) , \qquad (5.15)$$

assuming that $(H_{ac})_{max} = H_{dc}$. These relationships are not very well obeyed by practical tape powders since it is not possible to neglect particle interactions. A simple interaction model (see Ch. 2, § 5), involving negative interaction between pairs of particles with equal values of I_s, leads to relationships in better agreement with practical powders (SHTRIKMAN and TREVES [1960]). Despite its limitations, the formal interaction model assumed for the Preisach diagram gives relationships in agreement with the particle pairs model (BATE [1962]). The three modes of remanent magnetization acquisition are shown in Preisach diagram form in Fig. 5.12. Here, the diagram is divided into three areas A, B, C with the polarities of their magnetizations denoted for each method of magnetization. The equivalent relationships to eqs.

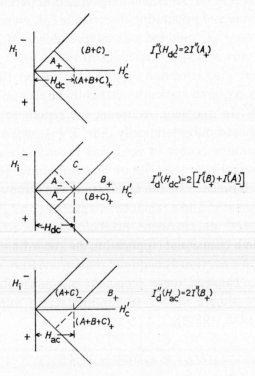

Fig. 5.12. Preisach diagrams for different modes of magnetization.

(5.14) and (5.15), taking interaction into account, are therefore given by

$$I_d''(H_{dc}) = I_r''(\infty) - [2I_r''(H_{dc}) + I''(c)], \qquad (5.16)$$

and

$$I_d''(H_{ac}) = I_r''(\infty) - [I_r''(H_{dc}) + I''(c)]. \qquad (5.17)$$

Without interaction effects, all particles would lie on the abscissae of the Preisach diagrams and areas such as $I''(c)$ are ignored. Using measured distributions of $I(H_i, H_c')$ for γ Fe$_2$O$_3$, good agreement is obtained with experimental curves for $I_d''(H_{dc})$ and $I_d''(H_{ac})$. Thus, both interaction models considered are able to account for the discrepancies in the initial relationships of eqs. (5.14) and (5.15).

The desirable remanent magnetization characteristic for linear response and high output in biassed recording techniques is one which rises linearly from zero at some critical field strength. The non-linear increase obtained in practice is similar for the different methods of remanent magnetization acquisition and depends on the distribution of particle switching fields. The effective switching fields are due to individual particle coercivities and interaction fields. Although the anhys-

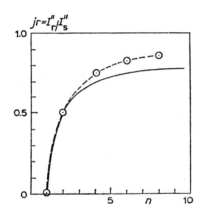

Fig. 5.13.   Reduced remanent magnetization *vs.* number of easy directions (*n*).
Random orientation. (WOHLFARTH and TONGE [1957]).
○– – – –○ spatial distribution; ——— planar distribution.

teretic remanent magnetization characteristic is dependent on the interaction fields alone, all of the remanent magnetization characteristics reflect the distribution of interaction fields and thus possess some degree of similarity.

§ 2.3.   *Magnetization Models for Single Domain Particles.* Non-Uniaxial Anisotropy

Although many magnetic powders used in magnetic tape consist of particles with uniaxial anisotropy, commonly due to shape anisotropy, commercial application of powders with non-uniaxial (cubic) anisotropy has also been successfully accomplished. This has been achieved by minimizing shape anisotropy through the use of spherical or cubical shapes in particles having high cubic crystal anisotropy. As will be shown, such powders have unique advantages for tapes: their magnetic properties are inherently size independent and very small particles with uniform properties are obtainable. Specific disadvantages, such as temperature dependence of magnetic properties, have, however, limited their application.

Two forms of non-uniaxial crystal anisotropy are of practical interest; cubic with six or eight easy directions of magnetization in a crystal, and planar crystal anisotropy with a number of easy directions equally distributed in a plane. The reduced remanent magnetization $(j_r = I_r''/I_s'')$ increases with the number of easy directions as shown in Fig. 5.13 (WOHLFARTH and TONGE [1957]) for planar and spatial distributions of the easy directions. Thus, it can be seen that much larger values of remanent magnetization are possible in such assemblies compared to $j_r = 0.5$, for $n = 2$ in uniaxial anisotropy particles. For instance, certain hexagonal ferrites have six easy directions in the basal plane. For an assembly of such particles with orientation of the hexagonal axes,

$$j_r = \int_{-\pi/n}^{+\pi/n} \cos\theta \, d\theta \bigg/ \int_{-\pi/n}^{+\pi/n} d\theta$$
$$= 0.96 \, (n = 6), \tag{5.18}$$

where θ is the angle between applied field (in the basal plane) and nearest easy axes.

For random assemblies of cubic anisotropy crystals with six or eight spatially distributed easy directions (e.g. iron and nickel, respectively) $j_r = 0.83$ and 0.86, respectively. Herein lies a property of great use in tapes where high output is required for different recording directions in the plane of the tape. To realize such properties it would, however, be necessary to suppress any forms of uniaxial anisotropy, which tend to reduce j_r to 0.5.

Since, in single domain spherical particles, the coercivity is determined by the crystal anisotropy, it is theoretically possible to achieve higher coercivities than may be obtained from shape anisotropy controlled particles. For larger particles some curling can take place between saturation and remanence, thus lowering the value of j_r. This condition is to be avoided in powders suitable for magnetic tapes. The anisotropy energy density for a cubic crystal depends on the magnetization direction (GANS [1932]) according to the relation

$$E'_c = K_0 + K_1(\alpha_1^2\alpha_2^2 + \alpha_2^2\alpha_3^2 + \alpha_3^2\alpha_1^2) + K_2\alpha_1^2\alpha_2^2\alpha_3^2 , \qquad (5.19)$$

where α_1 α_2 α_3 are the direction cosines of the magnetization vector with respect to the easy [001] axes. The total energy density is given by

$$E'_t = E'_c + E'_i , \qquad (5.20)$$

where E'_t is the total energy density, and E'_i the field energy density.

Assuming the applied field is in the opposite direction to the zero field magnetization direction, a maximum field energy is required to produce rotation, which is then equal to the maximum coercivity. If the rotation takes place in the (001) plane, eq. (5.19) is simplified to

$$E'_c = K_1(1 - \cos 4\theta)/8 \qquad (5.21)$$
$$\approx K_1\theta^2 \text{ for small values of } \theta .$$

Thus

$$E'_t = K_1\theta^2 - HI_s \cos(\pi - \theta) ; \qquad (5.22)$$

the K_2 term of eq. (5.19) is here neglected. The equilibrium condition is then

$$2K_1\theta - HI_s \sin \theta = 0 . \qquad (5.23)$$

The maximum energy will be required to produce a small initial rotation giving the maximum coercive force or anisotropy field:

$$H_{c(max)} = H_a = 2 K_1/I_s . \qquad (5.24)$$

For a random orientation of particles containing crystals with cubic anisotropy the coercive force is given by (NÉEL [1947])

$$H_c = 0.64 K_1/I_s . \qquad (5.25)$$

The very high values of H_a predicted for single crystals with large crystal anisotropies have been observed only in samples with no local shape anisotropy due to sharp corners. Comparisons between the measured coercive force and anisotropy field are tabulated for permanent magnet materials in Tables 6.1 and 6.2 of Chapter 6. Some results (denoted r) refer to a random orientation of crystallites, when eqs. (5.25) and (5.12) apply for cubic and uniaxial anisotropies respectively. For these two cases $H_c = H_a/3.1$ and $H_c = H_a/2.1$ respectively. Particle interaction effects may also lead to a reduction of H_c. For instance, it has been reported that H_c is constant up to a powder volume percentage of 46% in cobalt-iron oxide and falls for denser material (GUILLAUD [1953]).

The hysteresis curves for particles with cubic anisotropy have been calculated for an arbitrary angle between the applied field and the cubic axes (JOHNSON and BROWN [1961]). Putting eq. (5.19) in reduced form gives

$$e_t = (\text{sign of } K_1)(\alpha_1^2\alpha_2^2 + \alpha_2^2\alpha_3^2 + \alpha_3^2\alpha_1^2) . \tag{5.26}$$

Using an electronic computer, e_t may be calculated as a function of the direction of magnetization and the minimum energy equilibrium positions found. The resulting hysteresis loops are similar in shape to those obtained for uniaxial anisotropy, although some stable intermediate states are predicted which are not observed in practice. The previous calculations of I_r''/I_s'' for cubic anisotropy are confirmed.

§ 2.4.   Change of Magnetization with Time

Another aspect of the magnetization characteristics of single domain ferromagnetic particles of interest in magnetic recording is the time stability of magnetization. Potentially, changes of magnetization with time are harmful in two ways: in the decrease of recorded magnetization with time, and in the increase of magnetization due to fields from adjacent layers in a wound reel of tape. The basic phenomenon involved is magnetic viscosity due to thermal fluctuations.

When a magnetic field is applied to a material consisting of single domain particles, the resulting change in magnetization may take a finite time due to relaxation effects. For the same reason, when the field is removed, the magnetization may gradually decrease with time. The magnitude of this phenomenon depends critically on the particle volume and the temperature. In general, if the volume is small enough to be in the range where single domain properties give way to super-paramagnetic properties, then applied fields smaller than those necessary to switch the single domain particles can, in time, produce further magnetization rotation towards the field direction. The probability of magnetization in the field direction, in such circumstances, depends on the occurrence of favourable energy changes due to ambient temperature fluctuations. This means that the temperature fluctuations correspond to an energy of the same order of magnitude as that required to switch the magnetization of the particle from one easy direction to another. Since the lowest energy change corresponds to switching the magnetization towards the field direction, the probability of magnetization in the field direction depends also on the applied field magnitude, H. For instance, considering uniaxial, independent particles, the probability of their magnetization in the applied field direction in a time interval, dt, is given by P_τ, where

$$\begin{aligned} P_\tau &= dt/\tau \\ &= A \, dt \exp(- B/kT) \end{aligned} \qquad (5.27)$$

here τ is the relaxation time, B is the energy barrier corresponding to irreversible particle magnetization rotation, A is a frequency term depending on the applied field and the magnetization characteristics of the material, and k is Boltzmann's constant.

It is required, then, to investigate the physical factors which contribute to the relaxation time. If this time can be kept very long, then unwanted magnetization in magnetic tapes, due to extraneous small magnetic fields, will be minimized.

The parameter B may be readily evaluated. Considering a simple example of an applied field along the easy direction of the particles, the energy barriers to rotations towards and away from the applied field are given respectively by

$$\begin{aligned} B_1 &= I_s v (H_c - H)^2/2H_c, \\ B_2 &= I_s v (H_c + H)^2/2H_c. \end{aligned} \qquad (5.28)$$

These equations follow from the difference between the maximum and minimum particle energies during the magnetization rotation. The relevant particle energies may be obtained from eqs. (5.3), (5.9) and (5.10).

Evaluation of the frequency term A on the other hand is extremely difficult. Fortunately, however, different methods lead to the same primary dependence of the resulting relaxation time on the volume of the particles and the temperature, due to the exponential term in eq. (5.27) (BROWN [1959b], NÉEL [1949]). For instance, in the case of zero applied field, $B_1 = B_2$, and the probability of magnetization change is given by the sum of the now equal probabilities for change in either direction. The corresponding relaxation time τ_0 is given by

$$1/\tau_0 = 2A \exp(- I_s v H_c/2kT). \qquad (5.29)$$

One simple estimate of the frequency term A may be obtained from the theory of Brownian motion which assumes that A is equal to the natural frequency of oscillation of the system about the position of minimum potential energy (BROWN [1956b]). In magnetic materials the natural oscillation is due to the precession of the electron spins caused by gyromagnetic effects. Equating A to this frequency gives

$$A = eH_c/2\pi m \qquad (5.30)$$

where e and m are the electron charge and mass respectively, and an internal field of the order of H_c is assumed. Substituting in eq. (5.29) leads to relaxation times in good agreement with those calculated by more sophisticated approaches. Using the above expression for the relaxation time and applying typical values for I_s and H_c for acicular particles of iron oxide and iron (Tables 6.1 and 6.2) it is found that the particle volumes and relaxation times are as shown in Table 5.1 at room temperature.

It is to be noted that a change in particle volume of 2:1 causes a change in relaxation time of 10^9. Thus the particle volumes contributing relaxation times of significance in the stability of magnetic recordings or print-through are in the range of 1.5×10^{-18} cm^3 and 3×10^{-17} cm^3, respectively, for iron and iron oxide particles. Smaller particles will not contribute to the usable recorded magnetization and larger particles can be considered to be stable.

The above discussion can only be considered to illustrate time changes in magnetization of interest in magnetic recording. Any complete discussion would certainly have to include the detailed effects of particle interactions, distributions of relaxation times, and also time effects in particles too large to be truly single domains.

TABLE 5.1

τ_0 (sec)	v (cm^3)	
	Fe ($\times$ 10^{-18})	γ Fe$_2$O$_3$ ($\times$ 10^{-17})
10^{-1}	0.9	1.8
10^1	1.1	2.3
10^3	1.4	2.7
10^5	1.6	3.2
10^7	1.8	3.7

Print-through effects can be understood by extending the results of this discussion as described in Chapter 4, § 7.

§ 2.5.   *Application to Practical Powders*

The magnetization of small particles has been considered in this section and various magnetization modes have been described along with different controlling anisotropies. It has been shown that the predominant mode of magnetization change depends on the particle size and, in practical powders with a distribution of sizes, it may be expected that several modes coexist in a single sample.

From the point of view of magnetic tape properties where it is desirable to maintain a high ratio I_r''/I_s'', it is then necessary to constrain the distribution of particle sizes to avoid either superparamagnetic (sp) or multidomain (md) behaviour. As shown in Fig. 5.14 (BEAN [1955]), mixtures of (sp) and single domain (sd) particles, or of (sd) and (md) particles, have lower values of I_r''/I_s'' than for (sd) particles alone. On the other hand, if the spread of particle sizes is only sufficient to allow single domain behaviour, high values of I_r''/I_s'' are obtainable with either coherent or incoherent rotation mechanisms. For the coherent rotation mechanism, the critical size for single domain behaviour will increase with shape anisotropy whereas for incoherent rotation the particle width is the controlling factor.

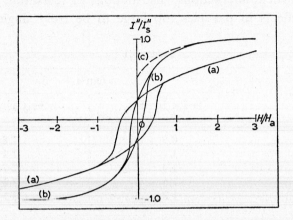

Fig. 5.14.   Hysteresis loops for mixed magnetization modes.
(a) Equal quantities single domain and multi-domain particles.
(b) Equal quantities single domain and paramagnetic particles.
(c) Single domain particles only.
       (BEAN [1955]).

Magnetocrystalline anisotropy on the other hand, being a local effect, has no influence on the critical size. It is still possible for a single domain sample, in which both magnetization mechanisms occur, to have a low value of I_r''/I_s'' if interaction takes place between the particles. An instance of this has been described in the particle pairs model (Ch. 2, § 5).

In addition to a high value of remanent magnetization, a rapid linear rise of I_r'' is desirable above some critical field. For particles controlled by shape anisotropy, it is then necessary to restrict the shape variations and thus minimize the spread of particle magnetization switching fields. Effective shape variations may also occur if the particles are porous. Most departures from the ideal described lead to a lowering of the overall coercivity. This in itself may not be detrimental for tape powders, but it is important to avoid a wide distribution of particle coercivities and interaction fields. The latter will depend on the uniformity of distribution of particles in the tape coating.

Another possible source of undesirable magnetization characteristics in small particles can occur if a mixture of anisotropies is operative in the same material. It has been shown (TONGE and WOHLFARTH [1958]) that if a uniaxial shape anisotropy is present with a non-uniaxial crystal anisotropy of comparable magnitude then the former is overriding with respect to the remanent magnetization, and for a non-oriented array, $I_r''/I_s'' \to 0.5$.

The coercive force would be increased or decreased depending on the relative orientations of the two anisotropies. The search for materials suitable for magnetic tapes would seem, then, to centre around the requirement that an easy and stable direction of magnetization exists near the applied field direction for each particle, and that the critical fields for irreversible magnetization change be the same for all particles. The properties of magnetic powders and films in these respects are reviewed in Chapter 6.

CHAPTER 6

MAGNETIC TAPE – PREPARATION AND PROPERTIES

§ 1.   MAGNETIC MATERIALS FOR TAPE

When the specification of the desirable magnetic properties for a tape material, obtained from the analysis of the recording and reproducing process, is considered with respect to the theoretical mechanisms of magnetization in permanent magnet materials it appears, at first sight, that a number of materials might be competitively considered for use in tapes. For instance, different types of anisotropy can yield similar magnetization characteristics, and thin magnetic layers can be produced in a variety of ways. However, the state of affairs, at the time of writing, is such that nearly all current commercial magnetic tapes have identical constructions and are fabricated from identical materials. The general construction consists of a plastic tape coated with a magnetic powder embedded in a plastic binder. With remarkably few exceptions the magnetic material used is gamma ferric oxide in the form of needle shaped (acicular) particles. It seems unlikely, on the other hand, that this unique preference can persist in the future. The commercial permanent magnet industry, with some related requirements to the tape industry, is currently producing a variety of attractive materials including metal alloys, metal powders, and oxide powders. Some of these materials will be described here with regard to their possible application to magnetic tapes.

The physical and magnetic properties of some permanent magnet oxide powders are collected in Table 6.1. Similar properties for thin films and powders of permanent magnet alloys are collected in Table 6.2. The magnetic properties of these materials are highly dependent on the final distribution of shape and size of the ferromagnetic regions in them, and consequently a range of properties can be obtained for a

176

TABLE 6.1

Physical and magnetic properties of some permanent magnet oxide powders

Material	Composition	Physical properties					Magnetic properties									Remarks
		Form	Crystal structure	Density ρ'	Length $a(\mu)$	Width $b(\mu)$	H_c Oe	σ_s gauss cm³/g	I_s'' gauss	I_r' gauss	H_{at} Oe	$K_1^{(1)}$ erg/cm³	H_a Oe	T_c °C	Reference	$K_1^{(1)} = K_1 \times 10^{-6}$ $I_s'' = 0.4 I_s$
Iron oxide	γFe_2O_3	P	Inverse spinel	4.98	1.0	0.2	250	80	160	120(e)	2000	0.047	2500(a)	675	Osmond [1952, 1953]	
		P	Inverse spinel	4.98	0.2	0.2	90	80	160	44(f)	—	0.047	235(b)	675		
	Fe_3O_4	P	Inverse spinel	5.21	0.2	0.2	115	92	192	52(f)	245	0.11	460(b)	585	Smit and Wijn [1959]	
Cobalt-iron oxide	$CoFe_2O_4$	P	Inverse spinel	5.29	—	—	4200	80	170	120(g)	1400	2.5	12000(b)	520	Ugine [1948]	
	$Co_xFe_{3-x}O_4$	P	Inverse spinel	5.00	0.08	0.08	640	73	146	100(g)	1300	1.0	5500(b)	—		$x = 0.15$
Barium ferrite	$BaFe_{12}O_{19}$	P	Hexagonal close packed	5.30	1.0	0.1	4000	72	152	114(e)	17500	3.0	17000(b)	450	Mones and Banks[1958] Smit and Wijn [1959]	Platelike particles
Barium-iron oxide with titanium-cobalt	$BaCo_xTi_x Fe_{12-2x}O_{19}$	P	Hexagonal close packed	5.34	1.0	0.1	1900	60	128	96(e)	—	1.3	8200(b)	—	Smit and Wijn [1959]	$\sigma = 0.5$
Lead ferrite	$Pb_{1.85}Fe_{11.5}O_{19}$	P	Hexagonal close packed	5.50	—	—	500	49	110	82(e)	—	—	—	—	Kojima [1956]	
Chromium oxide	$CrO_2 + Sb$	P	Tetragonal	—	1.0	0.1	380	90	—	—	—	—	—	130	Ingraham and Swoboda [1960]	
Manganese ferrite	$MnFe_2O_4$	P	Inverse spinel	5.00	—	—	—	80	160	—	—	—	100(b)	300	Smit and Wijn [1959]	
Nickel ferrite	$NiFe_2O_4$	P	Inverse spinel	5.33	0.06	0.06	123	49	108	—	—	—	412(b)	585	Berkowitz and Schuele [1959]	

(a) $H_a = 2\pi I_s$.
(b) $H_a = 2K_1/I_s$.

(e) Estimated value for partially oriented particles ($I_r''/I_s'' = 0.75$); assumed volume packing factor = 0.4.

(f) Similar to (e) with $I_r'/I_s'' = 0.27$.
(g) $I_r/I_s = 0.7$.

H_{at} Anisotropy field at 150 °C.
P Powder form.

TABLE 6.2

Physical and magnetic properties of some metals and alloys

Material	Composition	Physical properties					Magnetic properties								Reference	Remarks $K_1^{(1)} = K_1 \times 10^{-6}$
		Crystal structure	Form	Density ρ'	Length $a(\mu)$	Width $b(\mu)$	H_c Oe	σ_s gauss cm³/g	I_s^* gauss	I_r^* gauss	H_{sl} Oe	$K_1^{(1)}$ erg/cm³	H_s Oe	T_c °C		
Iron	Fe	body-centred cubic	P	7.88	0.04	0.015	825	218	680	460	9 800	0.4	10 000(a)	770	Luborsky et al. [1957]	
			T	7.88	—	—	400	218	1700(h)	1350(h)				770	Bozorth [1956]	
Cobalt	Co	hexagonal close packed	P	8.9	0.04	0.015	2100(r)	157	560	—	2 000	4.1	5 500	1120	Meiklejohn [1953]	
Nickel	Ni	face-centred cubic	P	8.9	—	—	—	54	194	—	90	−0.05	206	358		
Iron-cobalt	60Fe 40 Co	body-centred cubic	P	8.1	0.1	0.02	1075	235	750	565(e)	—	0.04	12 000(a)	1000(c)	Luborsky et al. [1957]	
Cobalt-nickel	82 Co 18 Ni	hexagonal close packed	T	8.9			250	143	1230	800					Koretzky [1963a]	
Cobalt-phosphorus	98 Co 2 P*	hexagonal	T	8.9	2μ layer		400	73	650	440					Sallo and Carr [1962]	Lamellar structure
Cobalt-nickel-phosphorus	75 Co 23 Ni 2P*	hexagonal	T	8.9	—		750	95	840	630					Bonn and Wendell [1953]	Lamellar structure
Iron-cobalt-nickel	55 Fe 40 Co 5Ni	body-centred cubic	P	—	0.1	0.05	760	—	314(i)	239					Iwasaki and Nagai [1962]	
Iron-nickel-chromium	76Fe, 12Ni, 12Cr	body-centred cubic	T	—	—	—	250	—	160	120				100	Hobson, Chatt and Osmond [1948]	
	74Fe, 8Ni, 18Cr	face-centred cubic	T	—	—	—	200	—	—	240				—		
Vicalloy II	13V, 35Fe, 52Co	body-centred cubic face-centred cubic	T	8.2	—	—	500			730(d)					Underhill [1948]	
Cunife I	60Cu, 20Ni, 20Fe	face-centred cubic	T	8.1	—	—	590			450					Underhill [1948]	
Cunife II	50Cu, 20Ni 2.5Co, 27.5Fe	face-centred cubic	T	8.6	—	—	260			580(d)					Underhill [1948]	
Cunico I	50Cu, 21Ni, 29Co	face-centred cubic	T	8.3	—	—	700			275					Underhill [1948]	
Cunico II	35Cu, 24Ni, 41Co	face-centred cubic	T	8.3	—	—	450			420					Underhill [1948]	

* Estimated nominal phosphorus content.
(a) $H_s = 2\pi I_s$.
(c) Virtual T_c.

(d) Directional properties.
(e) Estimated for partially oriented particles.
    $(I_r^*/I_s^* = 0.75)$: packing factor = 0.4.

P   Powder form.
T   Thin film.
(h) I_s^*, I_r^* (Theoretical values).

(i) $H_{max} = 2500$.
(r) Random orientation, 76° K.

given material. Thus, the values reported in the tables may be taken only as examples of what might be achieved in tapes manufactured from the materials listed. Structure – independent properties (density (ρ'), saturation specific magnetization (σ_s), Curie temperature (T_C), etc.) are also included in the tables and pertain to the properties of the bulk material.

§ 1.1. *Iron Oxide Powders*

Structure and Preparation

The ferromagnetic iron oxides of interest for magnetic tapes are magnetite (Fe_3O_4) and gamma ferric oxide (γFe_2O_3). The latter oxide may be produced by heating magnetite in an oxidizing atmosphere; it is unstable and transforms to an antiferromagnetic form (αFe_2O_3) on heating above 400°C. As described in Chapter 5, § 2.1, the spontaneous magnetization of magnetite is due to its divalent iron ions ($Fe_3O_4 \equiv Fe^{++}$. $Fe_2^{+++}O_4$) located in the octahedral spaces of the oxygen ion lattice. In order to denote the metal ion positions in the formulae for ferrites it is common practice to bracket those ions occurring in the octahedral sites. Thus

$$Fe_3O_4 \equiv Fe^{+++} \left[Fe^{++} \ Fe^{+++} \right] O_4 \ .$$

On oxidation to Fe_2O_3 the oxygen lattice remains unchanged and therefore the number of iron ions per unit cell decreases. The unit cell for magnetite contains eight molecules. On heating, transformation to ferric oxide occurs and the number of iron ions is reduced to $\frac{8}{9}$ of the number in magnetite, and a defect spinel structure is obtained. Thus for the unit cell

$$Fe_8^{+++} \left[Fe_8^{++} \ Fe_8^{+++} \right] O_{32} \rightarrow Fe_8^{+++} \left[Fe_{13\frac{1}{3}}^{+++} \right] O_{32} \ .$$

It is now believed that the vacancies are distributed in an ordered fashion, forming a superlattice. This may be indicated in the formula by enclosing the superlattice component in round brackets and denoting a vacancy by $\square$ (VAN OOSTERHOUT and ROOIJMANS [1958]). Thus the unit cell for $\gamma \ Fe_2O_3$ is $Fe_8^{+++}[(Fe_{1\frac{1}{3}}^{+++} \cdot \square \ 2\frac{2}{3}) \ Fe_{12}^{+++}] \ O_{32}.$

Since divalent iron ions have a magnetic moment of 4 Bohr magnetons per ion, the magnetic moment of magnetite is 4 Bohr magnetons per molecule. For γ Fe_2O_3 the magnetic moment is then 2.5 Bohr magnetons per molecule $Fe_2O_3(Fe^{+++} \equiv 5$ Bohr magnetons).

Magnetite powders may be produced in two ways, leading to distinctly different magnetic characteristics; this is due to their different particle shapes. One method produces approximately spherical particles whose properties are controlled by crystal anisotropy, and the other produces particles whose properties are essentially controlled by their shape. Since, on oxidation to γ Fe_2O_3, the size and shape of the particles remains essentially unchanged, the same general relations are preserved between the two types of particles. The magnetic anisotropy due to shape (length/width $\approx$ 5/1) considerably exceeds that due to crystal anisotropy ($K_1 \approx 10^5$ erg/cc. for Fe_3O_4), making the former controlling energy more suitable for tape powders. Single-domain acicular particles of γ Fe_2O_3, oriented in the direction of motion of the tape, are the magnetic constituent of practically all commercial tapes in use at present. However, since the crystal anisotropy of iron oxide may be considerably increased by addition of foreign metal ions (e.g. cobalt, § 1.2) the preparation of spherical particles will be considered here in addition to acicular particles.

Acicular particles of magnetite may be prepared by reducing red ferric oxide (α Fe_2O_3) or yellow hydrated ferric oxide (α FeO.OH) to the black magnetite at about 400°C. The pigment powders used for this purpose typically have lengths somewhat less than one micron, and length-to-width ratios of 5/1 to 10/1. This process is commonly written as follows (e.g. OSMOND [1953])

$$6\alpha \text{ FeO.OH} = 3\alpha \text{ Fe}_2\text{O}_3 + 3\text{H}_2\text{O} ,$$
and
$$3\alpha \text{ Fe}_2\text{O}_3 + \text{H}_2 = 2 \text{ Fe}_3\text{O}_4 + \text{H}_2\text{O} . \tag{6.1}$$

Careful oxidation at about 250°C produces γ Fe_2O_3

$$2 \text{ Fe}_3\text{O}_4 + \tfrac{1}{2}\text{O}_2 = 3\gamma \text{ Fe}_2\text{O}_3 . \tag{6.2}$$

The red oxide (α Fe_2O_3), used as the starting material, may be produced

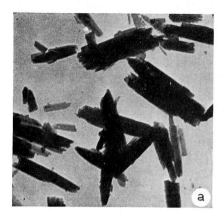

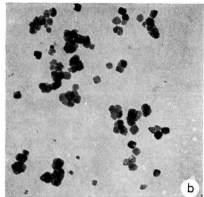

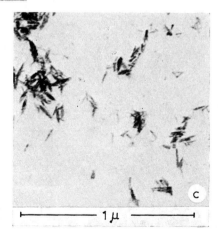

Fig. 6.1. Electron microscope
photographs of iron oxide particles.

in a variety of ways. Precipitation from a ferrous salt solution onto colloidal seed particles is a successful method, since the precipitation conditions can be controlled to produce particles of uniform prolate ellipsoidal shape and size (CAMRAS [1954]).

Approximately spherical particles of uniform size may be precipitated from a ferrous salt solution with, for instance, sodium or ammonium hydroxide. The resulting slurry may be oxidized, at about 80°C, to ferrous ferric oxide with ammonium nitrate which has an oxidation potential only sufficient to form Fe_3O_4. Air oxidation to γ Fe_2O_3 may then be performed as described above. The reactions are as follows:

$$
\begin{aligned}
FeSO_4 + 2NH_4OH &= Fe(OH)_2 + (NH_4)_2SO_4 , \\
3\,Fe(OH)_2 + \tfrac{1}{2}\,O_2 &= Fe_3O_4 + 3H_2O , \\
2\,Fe_3O_4 + \tfrac{1}{2}\,O_2 &= 3\gamma\,Fe_2O_3 .
\end{aligned}
\qquad (6.3)
$$

The particle sizes, and hence their coercivities, depend on the components of the precipitating solution and other experimental conditions. It is possible, in this way, to vary the particle size between 0.05 and 0.3 microns. By this method it is not possible, however, to produce sufficiently spherical particles for the crystal anisotropy to be dominant. Due to the low value of crystal anisotropy it would be necessary to reduce the effective length/width of the particles to less than 1.1/1 in order to render the shape anisotropy smaller than the crystal anisotropy (OSMOND [1954]).

Physical Properties and Critical Sizes

Many methods which have been devised for the production of ferrites are applicable to ferrous ferrite (Fe_3O_4). The first technique described above is an example of the method currently used in magnetic tape production producing acicular particles about 0.5–1.0 micron in length and a length/width ratio of 5/1 to 10/1 (Fig. 6.1a). The non-acicular particles produced by the second technique are about 0.1 micron in size (Fig. 6.1b). Modifications on these techniques can produce smaller acicular particles say 0.1 microns/0.02 microns as shown in Fig. 6.1c having similar magnetic properties to the larger particle but with the possibility of lower background noise level in magnetic recording (see Ch. 4).

Physically, it is expected that the acicular particles produced by

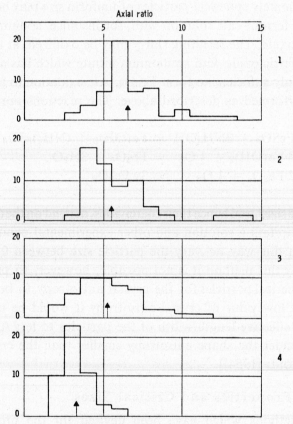

Fig. 6.2.   Particle shape distributions for acicular γ Fe₂O₃.
(Morrish and Watt [1957]).

reduction and reoxidation of αFe₂O₃ consist of a few single crystals with a common [110] axis parallel to the long axis of the particle. In practice it has been found that this is true (Van Oosterhout [1960]), although others have found that many low-order crystallographic axes are parallel to the long axis of the particle (Campbell [1957]).

Another departure from an ideal physical structure is evidenced by the occurrence of cavities in the particles. It can be assumed that the internal demagnetizing field structure will be modified by the presence of such cavities leading to a change of magnetic anisotropy. This modification of the ideal prolate ellipsoid may also have some influence on

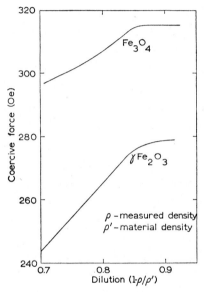

Fig. 6.3.   Dependence of coercive force on dilution of acicular iron oxide particles.
(WATT and MORRISH [1960]).

the mode of magnetization change, and favour the incoherent mechanism.

Typical distributions of particle shapes for acicular particles of γ Fe_2O_3 produced by reduction and reoxidation of αFe_2O_3, are shown in Fig. 6.2 for preparations with similar average particle lengths (MORRISH and WATT [1957]). For the coherent magnetization model, the shape distribution is responsible for a corresponding distribution of particle coercivities, and thus it would be important to maintain as narrow a range as possible. In the incoherent model, the particle coercivities are controlled by the minor axis dimension which, for these samples, will vary in the same way as the average shape. It was found that the concentration dependence of coercive force decreased from samples 1 to 4, suggesting a change from the probable interaction dependent mechanisms of fanning or buckling to multidomain behaviour which is independent of interaction. From the measurements of Morrish and Watt, the upper limit of the single domain range is in the region of 1000–2000 Å for the minor axis, for the 40 % volume packing density normally found in tape. The same authors (WATT and MORRISH [1960]) have also demonstrated that the single domain behaviour critical size for Fe_3O_4 is somewhat smaller than for $\gamma\,Fe_2O_3$ produced from it. Fig. 6.3 shows the interaction dependence of coercivity for these two powders, indicating a change to independence of packing density at 15 % volume packing density for Fe_3O_4, and at less than 10 % for $\gamma\,Fe_2O_3$.

The modes of magnetization change operating in sub-multidomain size particles of iron oxides have been largely determined from coercivity measurements. From Fig. 6.3 it is seen that acicular particles having a width of 1500 Å (length/width = 5) have intrinsic coercive forces less than 250 Oe and 300 Oe, respectively, for $\gamma\,Fe_2O_3$ and Fe_3O_4. A similar *isolated* $\gamma\,Fe_2O_3$ particle, on the other hand, has shown a coercive force of 800 Oe (MORRISH and YU [1956]), indicating the magnitude of interaction effects. It is also evident that a constant agglomeration occurs for dilute particle assemblies since the upper limits of coercive force in Fig. 6.3 are well below 800 Oe. Assuming coherent rotation, the anisotropy field for $\gamma\,Fe_2O_3$ is

$$H_a = 2\pi I_s = 2500 \text{ Oe}, \tag{6.4}$$

giving a measure of the maximum theoretical coercive force. Referring to Fig. 5.9 it is seen that the incoherent rotation mechanism requires the least field energy if the particle radius, S, is greater than the critical radius S_0. For $\gamma\,Fe_2O_3$, $S_0 = $ (exchange energy constant)$^{\frac{1}{2}}/I_s$ ≈ 250 Å. Thus, for the particles considered, $S/S_0 \approx 3$, giving a reduced coercive force of 0.1 (Fig. 5.10) for the curling mode; hence a coercivity of 250 Oe is predicted. This value is somewhat low compared to the single particle result. Also, the curling mode predicts independence of packing density, in contradiction to experiment. The experimental results could be better approached by considering the fanning mode to be operative in these powders. Further support for incoherent rotations in γ Fe$_2$O$_3$ has been given by a general agreement with the predicted angular dependence of coercivity shown for curling and

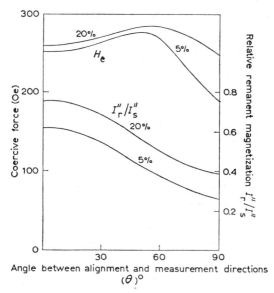

Fig. 6.4.   Coercive force and remanent magnetization *vs.* angle for aligned
γ Fe$_2$O$_3$ (BATE [1961]).

fanning in Fig. 5.10. Fig. 6.4 shows the angular dependence of H_c and I_r''/I_s'' for 5% and 20% volume packing fractions of γ Fe$_2$O$_3$ (BATE [1961]). The more pronounced H_c maximum for 5% packing is attributed to a greater degree of particle alignment.

The lower particle size limit for single domain behaviour is about 300 Å diameter for iron oxides. This agrees with experimental results which show that colloidal magnetite particles about 0.01 micron diameter are superparamagnetic (ELMORE [1938]). On the other hand, acicular particles with a minor axis of 0.04 micron are ferromagnetic, and this size may be guessed to be near the lower limit for stable ferromagnetic behaviour.

Magnetic measurements have been made of the distribution of critical fields for magnetization switching in acicular γ Fe$_2$O$_3$ powders, of the type currently used in commercial tapes, having a mean length-to-width ratio of about five (JOHNSON and BROWN [1958]). The remanent magnetization was measured after ac demagnetization of a previously saturated sample. The latter will be proportional to the number of particles whose critical fields exceed the maximum value of the ac field. Since the derived shape distribution was independent of temperature, it can be assumed that negligibly few superparamagnetic particles exist and that crystal anisotropy is also ineffective. Assuming a coherent magnetization mechanism, the particle shapes derived from analysis of the critical field distribution are about three times less acicular than those observed with an electron microscope, and it is more likely that incoherent processes predominate. Another magnetic measurement of the distribution of particle switching fields in acicular γ Fe$_2$O$_3$ involves only particles at approximately 90° to the applied field (FLANDERS and SHTRIKMAN [1962]). In this case (see Fig. 5.10), the coherent rotation mechanism is expected to predominate. Experimental confirmation is obtained, leading to closer agreement between the calculated particle shapes and those observed in the microscope.

Magnetization Characteristics

It has been shown that acicular particle assemblies of γ Fe$_2$O$_3$, having particle lengths around 0.5 micron and length-to-width ratios about five, behave like single domain arrays dominated by shape anisotropy. The oxide Fe$_3$O$_4$ behaves similarly and has a slightly larger saturation magnetization; it has not found application in tapes due to somewhat unstable room temperature characteristics. There is also a tendency for very small particles of Fe$_3$O$_4$ to oxidize at room temperature. Also some magnetic accommodation appears to take place which makes it difficult to permanently erase a recording on this material.

Acicular particles of γ Fe$_2$O$_3$ fulfill the general requirements for a magnetic tape powder. Since a single controlling uniaxial anisotropy operates in this material, it is possible to order this anisotropy (i.e. by particle orientation) to produce a high ratio of remanent to saturation magnetization. In addition, shape anisotropy depends only on particle geometry and saturation magnetization; consequently the magnetization is relatively stable against small temperature changes, since the Curie temperature (T_C) for this material is about 675° C. Preparation techniques have been described which produce particles with fairly good uniformity of shape and size leading to a corresponding uniformity of particle switching fields in a tape coating. Improvements in particle switching field uniformity would, however, be expected to produce improved short wavelength recording response of a tape. Finally, the ratio of remanent magnetization to coercivity is low enough in γ Fe_2O_3 to insure that self demagnetization losses are negligible. For the normal volume fraction of magnetic material in a tape (30–60%), demagnetization losses are small if this ratio is less than 0.3 for spherical particles, and less than about 4.0 for the acicular particles (see Ch. 5, § 2.1). This condition is easily met for γ Fe_2O_3, as can be seen from Table 6.1, and is, in fact, satisfied by all the acicular particles listed in the table.

The prime magnetization characteristics for various oxide powders are listed in Table 6.1 along with the material densities, (ρ') and typical particle shapes (denoted by major and minor axes a, b). Where possible, the coercivity (H_c) listed refers to the value obtained for a volume packing factor similar to that obtained in tape (assumed to be 40%). In some cases, only the values for highly compacted powders are available and these will, in general, be lower than the value expected in tape. The room temperature specific saturation magnetization (σ_s) gives a measure of the material magnetization independent of its porosity. Thus

$$\sigma_s = I_s/\rho'$$
$$= I_s'' v/a \tag{6.5}$$

where v is the sample volume, a the sample weight, and I_s and I_s'' are the material and sample saturation magnetizations, respectively.

At absolute zero temperature, the specific magnetization (σ_{s0}) may be related to the number of Bohr magnetons per molecule (n_B) in the

material, since
$$\sigma_{s0} = \text{magnetic moment/g} \tag{6.6}$$
$$= n_B\,\mu_B\,N/M$$

where N/M = number of molecules/g.

= Avogadro's number/molecular weight.

As shown earlier, for γ Fe$_2$O$_3$, $n_B = 2.5$.

The saturation and remanent intensities of magnetization (I''_r, I''_s) listed in Table 6.1 refer to typical values expected in tapes with 40% volume packing factor, and $I''_r/I''_s = 0.75$ or 0.7 for oriented acicular particles and random cubic crystals respectively. The dominant anisotropy is used to calculate the theoretical maximum coercivity, called the anisotropy field (H_a). When crystal anisotropy is dominant

$$H_a = 2K_1/I_s\,, \tag{6.7}$$

and when shape anisotropy is dominant

$$H_a = 2\pi I_s\,, \tag{6.8}$$

corresponding to the theoretical coercivity for an infinite cylinder assuming coherent magnetization rotation. The value of H_a at elevated temperatures (H_{at}) is also listed to give some idea of the increase in susceptibility to demagnetization.

The low values of coercivity for spherical particles of γ Fe$_2$O$_3$ and Fe$_3$O$_4$, listed in Table 6.1, are responsible for the large self demagnetization loss, giving $I''_r/I''_s \approx 0.3$ (see Fig. 5.3b). Hence, oriented acicular particles are of most interest, yielding a tape remanent magnetization (I''_r) of 120 gauss. Due to incoherent magnetization processes and interaction effects, the theoretical maximum coercivity of 2500 Oe is not even approached. The value of $H_c = 250$ quoted in the table is typical of acicular particles produced by the method described earlier. If the γ Fe$_2$O$_3$ so produced is taken through further cycles of chemical reduction and oxidation, the coercivity may be increased up to 400 Oe (CAMRAS [1954]).

Hysteresis loops for tape powders are the most frequently used characteristics in commercial evaluation. Along with curves of the rate of change of magnetization in sinusoidal cycling of the powder with an ac field, much useful information can be obtained with regard to the powders' suitability as a tape material. Families of loops, for a maximum applied ac field of 1000 Oe, are shown in Fig. 2.8 (Ch. 2), for oriented and non-oriented acicular particles of γ Fe$_2$O$_3$. The reduction

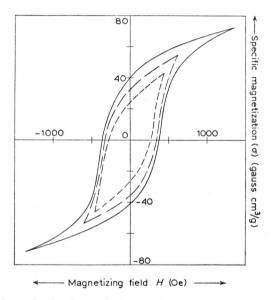

Fig. 6.5.   Magnetization loops for unoriented acicular particles of γ Fe$_2$O$_3$.
Volume packing factor = 0.2.

of reversible magnetization changes for magnetization in the direction of orientation is clearly seen with the associated increase of I_r''. For more precise measurements of hysteresis loops, magnetometer methods are commonly used for powder samples (§ 3.1). A typical set of loops for a dilute unoriented powder is shown in Fig. 6.5, measured with a vibrating sample magnetometer. It can be seen from the measured hysteresis loops that the remanent magnetization diminishes faster than the coercivity as the maximum field is reduced. The extent to which this is manifest is a measure of the uniformity of particle switching fields.

Remanent magnetization curves are of value for assessing the recording performance of a powder. Fig. 6.6 shows remanent magnetization curves for acicular particle tapes produced with and without an orienting field (curves (iii) and (ii), respectively), and for a random orientation of the same powder (curve (i)). It can be seen that some orientation is produced in sample (ii) by the process of laying the

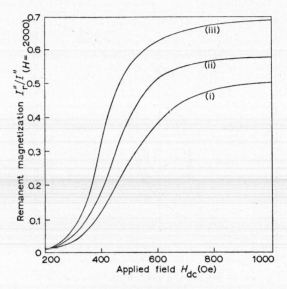

Fig. 6.6.   Remanent magnetization curves for γ Fe$_2$O$_3$.
(i)   Unoriented powder.
(ii) Tape, without orientation field.
(iii) Tape, with orientation field.

magnetic layer onto the plastic base material, giving intermediate maximum remanent magnetization. Orientation produces a steeper slope of the remanent magnetization curve reaching a maximum at a smaller field; this is due to all particles having similar modes of magnetization and similar critical fields for irreversible magnetization (see Ch 5, § 2.2). Orientation produces a somewhat similar change in the anhysteretic magnetization characteristic for γ Fe$_2$O$_3$, as can be seen in Fig. 6.7. As has been described, this characteristic is primarily dependent on the particle interaction fields and is not affected greatly by the individual isolated particle switching fields. The increased slope of the anhysteretic curve for oriented particles is in agreement with the theoretical discussion in Chapter 2, § 5.1.

The magnetic properties of iron oxides have been reviewed in this section with particular attention to γ Fe$_2$O$_3$, the almost universal

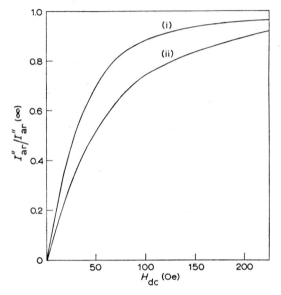

Fig. 6.7.   Anhysteretic magnetization curve for acicular γ Fe$_2$O$_3$ powder.
(i) Along direction of particle orientation.
(ii) Random orientation.

magnetic constituent of modern magnetic tapes. The more desirable magnetic properties of oriented acicular particles have been made evident for stable single domain particles of about 0.5 micron length. Due to the dominance of shape anisotropy, and the stability of saturation magnetization in the temperature range normally encountered in tape use, losses due to temperature variations are small. Smaller acicular particles have the advantage of lower background noise level although remagnetization during storage of a recording (print-through) is likely to increase. Other possible powders for tape will be considered in succeeding sections.

§ 1.2. *Cobalt-Iron Oxide Powders*

A possible alternative to acicular iron oxide powders exists in oxides with sufficiently high magnetocrystalline anisotropy to enable this to be the dominant anisotropy in a tape powder. In this case the particles may be isotropic in shape, and very small particles, not requiring orientation, can give the desired rectangular hysteresis loop. This approach has been used commercially for magnetic tapes (KRONES [1960]), by the addition of cobalt ions to cubic shaped iron oxide particles of about 0.1 micron size. Such additions produce non-stoichiometric cobalt ferrite having positive anisotropy (six easy directions) which increases with the cobalt content.

Cobalt-doped iron oxide may be prepared by the method described for non-acicular iron oxide with the addition of cobalt sulphate in the precipitating solution (AGFA [1954], JESCHKE [1954]). The starting solutions contain both divalent, and trivalent iron ions, and cobalt ions in an amount yielding up to 10 atomic per cent cobalt to total metal ions in the precipitate. Higher cobalt concentrations would give coercivities in excess of 1000 Oe due to the increase in magnetocrystalline anisotropy, and problems then occur in recording and erasing such a magnetically hard material. The large increase in magnetocrystalline anisotropy occurs both in cobalt substituted magnetite ($Co_xFe_{3-x}O_4$) and in the fully oxidized material, and is attributed directly to the cobalt ions on the assumption that they replace the divalent iron ions on some of the octahedral B sites (SLONCZEWSKI [1958]). Thus, the formula for cobalt substituted magnetite would be $Fe^{+++}(Co_x^{++} Fe_{1-x}^{++} Fe^{+++})O_4$.

The particles produced by the method of precipitation from the metal salts can have a large size range. When the starting solution contains both divalent and trivalent metal ions, a good uniformity of sizes is obtained in the range 600–1000 Å depending on the experimental conditions of the precipitation. As will be seen, such particles exhibit high ratios of remanent to saturation magnetization indicative of single domain behaviour controlled by non-uniaxial anisotropy. When the precipitating conditions favour a high nucleation rate for particles, but a slow rate of growth (ELMORE [1938]), particles may be obtained which are small enough to be superparamagnetic. It has been shown experimentally that the lower limit of crystallite size for ferromagnetic behaviour in cobalt ferrite is about 150 Å (SCHUELE and DEETSCREEK [1961]). Thus smaller particles than those described above (600 – 1000 Å) might provide tapes with lower background noise while retaining ferromagnetic stability.

The magnetic characteristics for particles of pure cobalt ferrite and for iron oxide doped with 5 atomic per cent cobalt are shown in Table 6.1. Cobalt ferrite is too hard magnetically to be considered for conventional magnetic tape use but has been included in the table to illustrate the exceptionally high anisotropy attainable in this material. The large coercivity of 4200 Oe reported for stoichiometric cobalt ferrite is similar to the theoretical value for a random array of such particles. From eq. 5.25

$$H_c = 0.64 \, K_1/I_s$$
$$= 4000 \text{ Oe at room temperature for } CoFe_2O_4.$$

For the lower cobalt content powder, small particles may be produced (800 Å) with a specific saturation magnetization some 10 % lower than $\gamma \, Fe_2O_3$ or cobalt ferrite, giving a similar reduction of tape magnetization. Due to their large crystal anisotropy the cobalt ions produce an increase of the coercivity of cobalt substituted iron oxide depending almost linearly on the cobalt concentration. A small atomic percentage of cobalt is sufficient to produce an adequate coercivity for tape with dominant control by crystal anisotropy.

An undesirable feature of crystal anisotropy is its large temperature dependence. The magnetocrystalline energy is related to the saturation magnetization although the functional relationship varies from one material to another. A consequence of this dependence is that the magnetocrystalline energy is strongly temperature dependent, and is zero at the Curie point. On the other hand, the saturation magnetization of cobalt ferrite is not so highly temperature dependent.

Consequently, moderate temperature rises cause a large change in anisotropy field and hence in coercivity (see H_{at}, Table 6.1). The dependence of anisotropy field $(2K_1/I_s)$ on temperature for $Co_{0.8}Fe_{2.2}O_4$ (26.7 atomic % Co) is shown in Fig. 6.8 (SUGIURA [1960]). In the curves shown, the uniaxial anisotropy induced by magnetic annealing

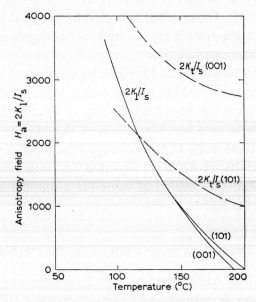

Fig. 6.8.   Temperature dependence of anisotropy field for $Co_{0.8}.Fe_{2.2}.O_4$.
K_1 – Crystal anisotropy constant.
K_t – Induced anisotropy constant (SUGIURA [1960]).

K_t, is also plotted, exhibiting a smaller temperature dependence. The reduction of coercivity with temperature leads to a loss of remanent magnetization of a recorded tape as shown in Fig. 5.3b, and this is one disadvantage of crystal anisotropy controlled particles in tape. The magnitude of this loss is indicated in Table 6.1 where H_{at} is shown to be 24% of H_a for an unannealed spherical powder sample with room temperature coercivity of 640 Oe.

Some stabilization may be achieved by magnetic annealing. By subjecting the material to sufficient simultaneous heat and magnetic field energy, diffusion of cobalt ions takes place to ordered positions which are determined by the direction of the applied field with respect to the crystal lattice. The effect is to introduce a uniaxial anisotropy (K_t) which will add to the existing crystal anisotropy causing an increase of coercivity. The temperature variation of the anisotropy field

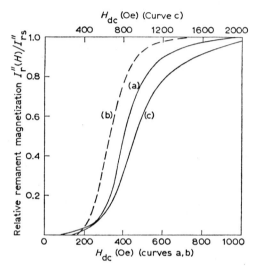

Fig. 6.9.   Relative remanent magnetization curves for
(a) $Co_{0.1} Fe_{2.9}O_4$
(b) $\gamma\ Fe_2O_3$ (oriented)
(c) Alloy powder.

($2K_t/I_s$) due to the induced uniaxial anisotropy is shown in Fig. 6.8 to be lower than that of the crystal anisotropy field.

Hysteresis loops for cobalt-doped iron oxide powders are shown in Chapter 2 (Fig. 2.10b). The measured loops are for a sample with about 3 atomic per cent of cobalt producing a coercivity of 300 Oe. In contrast with the $\gamma\ Fe_2O_3$ loops (Fig. 2.8), the same loop shape is obtained for all magnetization directions. However, the loop squareness is slightly inferior to the oriented $\gamma\ Fe_2O_3$ sample and $I_r''/I_s'' = 0.7$ (assuming $I_{1000}''/I_s'' = 0.9$) which is somewhat lower than the theoretical maximum value of 0.83 for this material (Fig. 5.13). This characteristic

is also illustrated by a smaller maximum slope in the remanent magnetization curve shown in Fig. 6.9, in which the maximum values have
been normalized. The greater influence of internal fields in the spherical
particle cobalt-doped iron oxides may also be seen in its lower initial
anhysteretic susceptibility (Fig. 6.10) which is in better agreement with

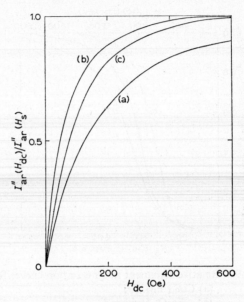

Fig. 6.10.   Anhysteretic remanent magnetization curves for
(a) $Co_{0.1} Fe_{2.9}O_4$.
(b) Oriented γFe_2O_3.
(c) Alloy powder.

the theoretical value calculated for random internal fields than that
obtained with γFe_2O_3. The alloy powder curves shown in Figs. 6.9 and
6.10 will be discussed in the next section.

In summary, an increase of the magnetocrystalline anisotropy of
iron oxide may be achieved by adding cobalt. Controlled permanent
magnetic properties may thereby be obtained in very small spherical
particles and the desirable features of a magnetic tape powder are
obtained. In comparison with acicular γFe_2O_3, high coercivities are

possible, and rectangular hysteresis loops are obtained regardless of the direction of magnetization. The thermal stability of magnetization is, however, inferior.

§ 1.3.   *Metallic Powders*

The noncompensated antiferromagnetism of ferrites is responsible for their relatively low saturation magnetizations compared to those of the ferromagnetic metals, iron and cobalt. The physical and magnetic properties of metals and alloys in the form of powders and thin films are listed in Table 6.2 where the alloy compositions are listed by weight percent. Here it may be seen that saturation magnetizations of iron and cobalt powders, diluted to 40% by volume for a tape coating are 680 and 560 gauss respectively, whereas all the oxides listed in Table 6.1 have tape saturation magnetizations less than 200. Nickel powder, on the other hand, has properties not unlike those obtained with γ Fe_2O_3. Thus, the metals iron and cobalt, their alloys with each other, and with nickel and other elements, have the theoretical possibility of yielding remanent magnetizations up to four times greater than oxide powders. Providing suitable coercivities can be developed along with rectangular hysteresis loops, the advantages of metals may be used either directly, by substituting metal for oxide powders in tape coatings, or by using thinner metallic layers such as may be obtained by evaporation in vacuum or by electro- or chemical deposition. Metal powders are discussed in this section and metal thin films in § 1.5.

Preparation and Physical Properties

Referring to Table 6.2 it is seen that the crystal anisotropy is some ten times greater for iron than γ Fe_2O_3, and the saturation magnetization rather more than four times greater. Thus, the anisotropy field will be about 500 Oe leading to $I_r''/H_c \approx 1$. This ratio is considered to be too high (§ 1.1) to avoid demagnetization losses. On the other hand, the theoretical anisotropy field, for shape anisotropy controlled iron particles, is $2\pi I_s$ ($= 10\,000$ Oe) assuming coherent rotation processes. These do not occur in acicular iron particles, however (e.g. LUBORSKY [1961]) which change their magnetization by an incoherent mechanism, probably fanning. As a consequence, maximum coercivities around 1 000 Oe are obtainable in practical shape anisotropy controlled iron particles. The coercive force is usually strongly dependent on particle size, and is a maximum for a particle diameter of 100–200 Å, as shown in Fig. 5.5, where single domain behaviour is expected (Fig. 5.4). However, the coercive force of highly elongated iron powders (length/ width = 10) appears to be independent of particle diameter over the range of 100–300 Å, in agreement with the fanning mode shown in Fig. 5.9 (LUBORSKY [1961]). Due to the large magnetic moment of iron, interaction between particles is of importance in determining the magnetic properties of powder compacts. Much effort has been directed towards the production of permanent magnets from elongated single domain particles of iron, and successful powders for magnets would appear to be eminently suitable for magnetic tapes.

One distinct hazard with fine powders of metals is their highly pyrophoric nature, requiring adequate protection against oxidation in a tape layer. There are numerous methods for preparing fine metal powders. Two general techniques will be considered here, firstly electrodeposition into mercury, and secondly reduction techniques.

Elongated particles may be produced by electrodeposition from an electrolyte of the metal salt, say ferrous sulphate, into a mercury cathode if mechanical agitation of the cathode is avoided (LUBORSKY et al, [1959]). Any vibration appears to prevent the growth of elongated particles. The deposition conditions are adjusted to form a minimum of unwanted dendritic growth on the elongated particles. By a critical heating of the particles in the mercury the small dendrites may be dissolved, leaving particles with about 150 Å diameter and length-to-width ratios of up to 10:1. The magnetic properties of typical particles are listed in Table 6.2. for a length-to-width ratio of 3:1 and a moderate degree of orientation, $I_r''/I_s'' = 0.65$. An electron microscope photograph of iron-cobalt particles produced by this method is shown in Fig. 6.11a (LUBORSKY [1961]). Since the iron particles are controlled by shape anisotropy, and since their saturation magnetization is high, particle interaction effects are strong. The consequent reduction of coercivity with increased packing density can be attenuated by improving the particle dispersion. This has been achieved by forming an amalgam between the mercury cathode and a nonmagnetic metal which becomes adsorbed on the iron particles and keeps them magnetically separated. The iron particles can be extracted magnetically from the mercury, final separation being achieved by vacuum distillation. This technique is equally suitable for cobalt, and iron-cobalt alloy particles. Deposition of cobalt from a cobalt sulphate bath produces hexagonal cobalt particles with a thin oxide layer: this layer does not significantly reduce the magnetization of the material but does render the particles nonpyrophoric. Iron-cobalt alloy particles have the advantage of higher saturation magnetization than either iron or cobalt, and, since shape anisotropy is dominant, a correspondingly higher coercivity. The magnetic characteristics of elongated particles of a 60:40 iron-cobalt alloy, listed in Table 6.2, show the expected increase of H_c and I_s''. This alloy has the highest saturation magnetization known.

Reduction of metal and alloy oxides may be carried out directly with hydrogen. For instance, acicular iron powders are produced on heating acicular γ Fe_2O_3 particles to about 250° C in an atmosphere of pure dry hydrogen. Higher temperatures have been used but excessive particle sintering then occurs. It is found that the resulting iron powders have reduced acicularity compared with the starting material. Coercivities of about 600 Oe are obtained for 250 Å diameter particles and 2:1 length-to-width ratio, with a 40% packing density (CARMAN [1959]). This is lower than the coercivity of electrodeposited particles with a 3:1 length-to-width ratio. Fig. 6.11b shows an electron microscope photograph of iron particles produced by this method. Reduction of iron oxide with calcium hydride at relatively low temperatures ($\approx$ 200° C) gives similar results with possible application to large scale production (CAMPBELL et al. [1957]). Somewhat larger acicular particles have been developed for magnetic tape use, (FEICK and STEDMAN [1960]) in which improved acicularity of the final product is claimed, through an initial heat treatment of the oxide, which may densify it and reduce break-up during reduction.

Metal oxalates have been used as the starting material for the preparation of fine particles of metals, alloys, metal oxides and ferrites

(SCHUELE [1959]). Since the oxalates are precipitated from a solution of the metal salts it is a simple matter to precipitate solid solutions of ferrous oxalate with the metal oxalates. Reduction of the oxalates to the metals or alloys may be carried out by heating in a hydrogen atmos-

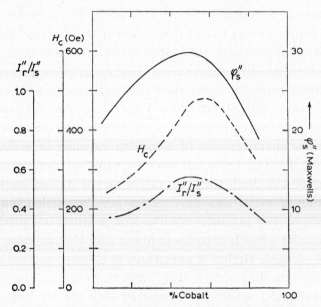

Fig. 6.12.   Magnetic characteristics of iron-cobalt alloy powders
(NAGAI et al. [1959]).
φ_s'' – Saturation flux (cf. $\varphi_s'' = 14$ maxwells for Fe_3O_4).
H_c – Coercive force.
I_r''/I_s'' – Ratio of remanent to saturation magnetization.

phere to a temperature high enough to effect decomposition but low enough to avoid sintering. Binary alloy powders of iron-cobalt have been produced successfully by this method and also ternary alloys of the iron-cobalt-nickel system (UGINE [1947], NAGAI et al. [1959, 1960a, b]). Less acicular particles of Fe-Co-Ni are shown in Fig. 6.11c produced from oxalates and the corresponding coercive force is lower than for a similar Fe-Co alloy as shown in Table 6.2. The variation of coercive force and saturation flux for unoriented powder samples of

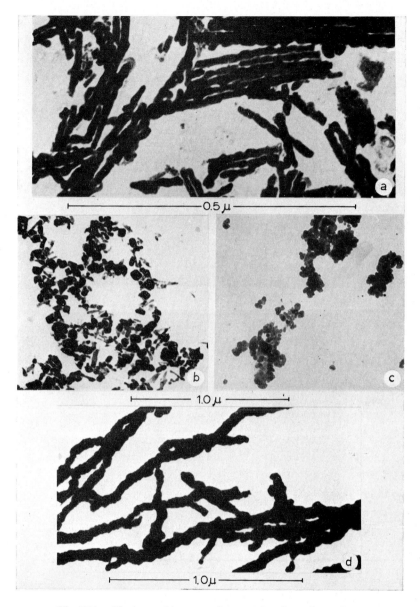

Fig. 6.11. Electron microscope photographs of metal powders.
a. Electrodeposited iron-cobalt.   c. Iron-cobalt-nickel by oxalate reduction.
b. Iron by oxide reduction.   d. Iron by borohydride reduction.

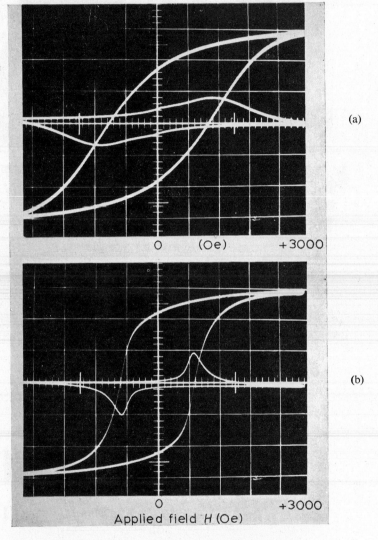

O   (Oe)   +3000

O   +3000

Applied field H (Oe)

Fig. 6.14.   60 c/s hysteresis loops and differential curves for:
       a. Unoriented iron powder tape. $H_c = 1000$ Oe.
       b. Oriented alloy powder tape. $H_c = 720$ Oe.

(a)

(b)

Fig. 6.16.   Electron microscope photographs of barium ferrite particles.
a. Domain patterns on basal plane of oriented compact (CRAIK and TEBBLE [1961]).
b. Small platelets.

iron-cobalt particles is shown in Fig. 6.12 as a function of cobalt content.

Finally, extremely acicular metal and alloy powders have been produced by reduction with borohydrides (OPPEGARD et al. [1961]). Particle lengths are increased if a magnetic field up to 1000 Oe is applied during the reduction, and iron particles may be 1–2 microns long and 0.04–0.1 microns in diameter Fig. 6.11d. For a volume packing density of about 40% the coercivity is in the range 500–700 Oe, somewhat higher than that obtained in reduced oxides. Single domain behaviour is therefore achieved in these relatively large particles, and incoherent magnetization reversal processes appear to take place.

Magnetic Characteristics

The prime magnetic characteristics for metal and alloy powders, suitable for magnetic tape application, are listed in Table 6.2. In most cases the properties refer to particles of optimum size for single domain behaviour. Smaller and larger particles would tend to be superparamagnetic or multidomain, respectively. In general, the particles are smaller than the oxide particles now used in tapes, and this has been shown to result in a lower background noise level when compared with similar but larger particles. The magnetic characteristics for iron, iron-cobalt and cobalt correspond to particles produced by electrodeposition, whereas the iron-cobalt-nickel results were obtained with particles produced by oxalate reduction. In the latter case the low value of I_s'' is possibly due to a low packing density. Nevertheless, for all the powders listed (apart from nickel), both H_c and I_s'' are considerably increased compared to the values obtained for iron oxides.

For the acicular metal and alloy powders described in Table 6.2, the magnetic properties are determined almost entirely by the particle shape and size. It has been shown (LUBORSKY et al. [1958]), that for elongated iron particles the crystal anisotropy contribution is negligible; for iron-cobalt particles it is also very small but the observed temperature dependence of coercivity suggests that it opposes the shape anisotropy. Due to the dominance of shape anisotropy and the large saturation magnetization of metals, strong interparticle interaction occurs leading to a strong dependence of coercivity on packing density. Experimental measurements of the dependence of coercive force of acicular iron and iron-cobalt particles on packing density show a linear dependence at low packing densities. At moderate packing densities, the iron-cobalt particles show a greater dependence

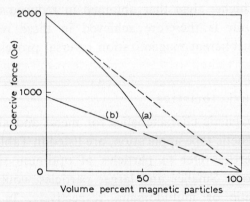

Fig. 6.13.   Dependence of coercive force on packing density
(a) Fe–Co (axial ratio 5.4 : 1) (LUBORSKY *et al.* [1959]).
(b) Fe (axial ratio 2 : 1) (CARMAN [1959]).

of coercivity on packing density (Fig. 6.13). The latter effect is thought to be due to a loss of single domain behaviour where particles touch.

60 c/s hysteresis loops for an unoriented acicular iron powder tape and an oriented alloy powder tape are shown in Fig. 6.14. In the former tape, saturation is barely reached for an applied field of 3000 Oe and the wide peaks of the differential loop indicate a considerable spread of particle switching fields. Better properties are shown by the alloy powder tape where, with a coercivity of 720 Oe, all irreversible changes are completed for a field of 1800 Oe. The loop shape is similar to that obtained with oriented oxide powders and, in fact, the magnetization processes are the same. It has been shown that magnetization occurs in acicular iron powders by an incoherent fanning process and support

for this is obtained in the angular dependence of coercive force for an oriented array which exhibits a maximum at 50–60 degrees (see Fig. 6.15); this is similar to the angular dependence of coercive force for iron oxide (Fig. 6.4). The remanent magnetization curve and anhysteretic magnetization curve are plotted in Figs. 6.9 and 6.10, respectively

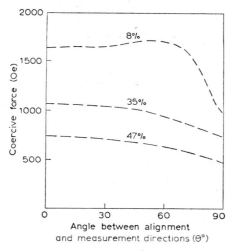

Fig. 6.15.   Dependence of coercive force of aligned iron particles on angle. 8% sample measured at − 196°C. 35% and 47% samples measured at room temperature (LUBORSKY and PAINE [1960]).

for the alloy powder tape. Similarity with the oxide curves is evident for the latter. For the remanent magnetization curves, noting the different scale of the abcissa, it appears that the relative spread of particle switching fields is somewhat larger for the alloy powder tape.

§ 1.4.   *Miscellaneous Oxide Powders.*
   Ferrites with Hexagonal Structure

When appropriate mixtures of iron oxide with barium, strontium or lead oxides are fired at high temperatures a hexagonal structured ferrite results with the formula unit of the type $BaFe_{12}O_{19}$. The detailed structure of the unit cell is a mixture of hexagonally close packed oxygen layers, in which some oxygen ions are replaced by barium, and cubic close packed oxygen layers as found in the spinel structure. Ten oxygen layers occur in a unit cell and the number of ions corresponds to $2(Ba^{++}Fe_{12}^{+++}O_{19})$. The saturation magnetization per formula unit is 20 Bohr magnetons yielding $\sigma_{s0} = 100$ gauss cm^3/g (from eq. 6.6). Experimental measurements of σ_s at low temperatures agree with this figure and the variation of σ_s with temperature is shown in curve (d) of

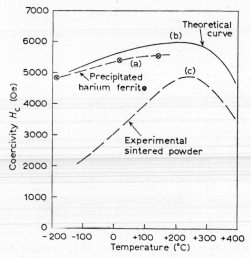

Fig. 6.17a.   Temperature dependence of coercivity for barium ferrite.

Fig. 5.1. The room temperature value of σ_s equals 72 gauss cm^3/g, which corresponds to $I_s = \rho' \sigma_s = 380$ gauss, using the values quoted for $BaFe_{12}O_{19}$ in Table 6.1. In practice, the very small single domain particles of barium ferrite have a saturation magnetization a little lower than the bulk material value, presumably due to a surface layer with different properties (TORKAR and FREDRIKSEN [1959]).

The critical size for single domain behaviour has been determined by direct observation, in the case of barium ferrite, by using a modification of the powder pattern technique. The colloidal magnetite indicator is allowed to dry in a stabilizing agent, then evaporated with carbon and viewed in an electron microscope. As can be seen from Fig. 6.16a (CRAIK and TEBBLE [1961]), small single domain particles,

having a large normal component of magnetization, collect correspondingly large areas of colloid. Multidomain particles, on the other
hand, show a domain pattern since the colloid is attracted to the domain boundaries only. It is observed that single domain particles have
a diameter of about 1 micron or less. Small single domain particles
(0.15 micron) may be prepared by precipitation techniques of the type
used to prepare small particles of iron oxides, and uniformly small
plates – haped particles are obtained (Fig. 6.16b). These particles have

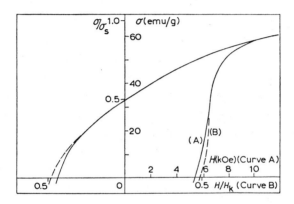

Fig. 6.17b.   Magnetization loop for unoriented barium ferrite.
              (A) Small precipitated particles.
              (B) Theoretical curve.

an extremely high room temperature coercivity of over 5000 Oe which
changes little over the temperature range −200°C to 150°C (Fig.
6.17a – curve (a)). Assuming a demagnetizing factor of 4π for the
platelets, and the easy magnetization axis to be normal to their plane,
the theoretical coercive force (eq. (5.12)) for a randomly oriented
sample is

$$H_c = 0.48 \left[(2K_1/I_s) - 4\pi I_s \right]. \tag{6.9}$$

The theoretical dependence of H_c on temperature for barium ferrite
obtained by inserting measured values of K and I_s is plotted in Fig.
6.17a – curve (b). Good agreement is obtained with the experimental
results. Furthermore, the experimental magnetization loop for this
material is in excellent agreement with eq. (5.11), derived from the

coherent rotation model, as shown in Fig. 6.17b (MEE and JESCHKE [1963]). The curves are normalized to the experimental value of $\sigma_s = 68.0$ for the sample.

Electrolytic coprecipitation methods can also be used to prepare barium ferrite particles (BEER and PLANER [1958]); as with the chemical precipitation method referred to above an intimate mixture of the constituent oxides is obtained requiring relatively low temperatures to produce the ferrite. In this way, sintering is avoided and small individual particles are obtained. Larger particles (Fig. 6.17a – curve (c)) exhibit single domain behaviour at elevated temperatures only where the critical size for single domains is increased. Due to the temperature stability of the coercive force of the small particles, irreversible losses of remanent magnetization do not occur on heating. From this point of view, barium ferrite is superior to cobalt-doped iron oxide; its major disadvantage for a magnetic tape powder is concerned with the very large coercive force occurring in single domain particles.

The coercive force of single domain particles may be reduced in several ways, but at the expense of deterioration of other desirable magnetic properties. As indicated, larger particles exhibit multidomain behaviour and thus lower coercive force. However, to achieve sufficiently low coercive force, a particle size of several microns is required and this is considered to be too large for low noise tapes (see Ch. 4).

Another very interesting set of hexagonal ferrite materials exhibits lower coercive force than the type $BaFe_{12}O_{19}$ due to the development of easy directions of magnetization in the plane perpendicular to the hexagonal c axis; the c axis itself then becomes a difficult direction (JONKER et al. [1957], STUIJTS and WIJN [1958]). These materials contain additional divalent ions such as cobalt and have six easy directions of magnetization in the basal plane. For small platelet shaped particles with easy axes in the plane of the plate, it would be possible to produce a magnetic tape with the plates parallel to the tape plane, giving easy magnetization in the plane of the tape ($I_r''/I_s'' = 0.96$, see Eq. (5.18). It would then be difficult to magnetize such a tape perpendicular to the tape plane, a condition which leads to high resolution recording. Finally, the crystal anisotropy of uniaxially anisotropic hexagonal ferrites may be reduced by replacing some of the ferric ions by divalent cobalt and tetravalent titanium ions in equal amounts, giving the formula unit $BaCo_x^{++}Ti_x^{++++}Fe_{12-2x}O_{19}$. For $x = 0.5$, for instance, the crystal anisotropy is reduced to 40% of the value for $BaFe_{12}O_{19}$, while the saturation magnetization reduces by only 17% (CASIMIR *et al.* [1959]), as indicated in Table 6.1. The coercivity may then be expected to fall by 53%. Although this value is still very large by normal tape standards, such a material might find limited use for special applications requiring high permanence of recordings.

Somewhat similar results are obtained for hexagonal ferrites containing strontium or lead rather than barium (KOJIMA [1958]) and the saturation magnetization may be further increased by adding a few percent SiO_2 or Bi_2O_3. Interesting properties are obtained with lead ferrite since a lower coercive force is obtained (about half that of barium ferrite) while the saturation magnetization is similar to other hexagonal ferrites. Most additionals to lead ferrite produce a small increase of I_r and H_c. Adding a few percent SnO, however, produced a reduction of H_c to around 500 Oe, while maintaining $I_r'' = 158$ gauss for an unoriented powder. The magnetic properties expected of oriented lead ferrite particles are given in Table 6.1.

Ferrites with Spinel Structure

In addition to the iron and cobalt ferrites already considered, manganese, nickel, copper, magnesium and lithium ferrites have simple spinel structures. The first two of these, having specific saturation magnetizations approaching those of iron and cobalt ferrites ($\sigma_s = 80$ and 50 respectively) are listed in Table 6.1. Manganese ferrite, having the largest saturation magnetization at low temperatures, has a low Curie point which is not conducive to stable properties around room temperature. Nickel ferrites are of some possible interest but the low values of I_s, and even lower for copper, magnesium and lithium ferrites, militate against their use in competition with iron oxides.

Oxides with Tetragonal Structure

Chromium dioxide, having a specific saturation magnetization in the range 80–100 gauss cm³/g, may be prepared in fine powder form by the thermal decomposition of chromium trioxide in water at high pressure (SWOBODA *et al.* [1961]). The addition of trace amounts of antimony or ruthenium substitutionally for chromium produces acicular particles of typical length-to-width ratios of 10:1; it also has the effect of reducing the temperature and pressure of the hydrothermal reaction during the preparation. The magnetic properties of such powders are listed in Table 6.1. It has been shown (*loc. cit.*) that the acicular particles are single crystals with the tetragonal axes parallel to the long axes of the particles. Somewhat superior magnetic properties are therefore achieved at room temperature compared to iron oxides. A possible disadvantage of this material exists in its rather low Curie point.

§ 1.5. *Metal Tapes*

Since the requirements for magnetic tape are a combination of magnetic hardness and physical softness, it is not easy to produce satisfactory results from all-metal tapes; in general, magnetic and physical hardness go hand in hand. Exceptions to this rule fall into two groups. Firstly, some magnetic alloys of iron-cobalt, nickel-iron-cobalt, and nickel-cobalt form ductile products when alloyed with copper or vanadium. These may be rolled to thin tapes suitable for magnetic recording when very high resolution is not required as, for instance, in recording digital information. Some of these alloys with the trade names 'Vicalloy', 'Cunife' and 'Cunico' are listed in Table 6.2. Magnetic stainless steel also falls into this category and has been used for wire recording. However, all of these materials suffer from physical rigidity, even in the form of thin tapes or wires, leading to imperfect contact with the recording/reproducing transducers and consequent separation losses (see Chapters 4 and 7). Secondly, if sufficiently thin layers (a few microns) of the ferromagnetic metals and alloys are formed, magnetic hardness increases due to the restriction of the direction of magnetization to the plane of the layer. Even greater magnetic hardness may be incorporated if the layer itself is not homogeneously ferromagnetic. If such thin layers of metals could be suitably

laid onto a plastic tape, then the advantages of high saturation magnetization alloys might be combined with high coercivity in a physical
form conducive to efficient magnetic recording.

The ductile magnetic alloys used for magnetic recording tapes or
wires are, in principle, quite similar magnetically to the acicular iron
oxides currently used in magnetic tape. That is to say, they consist of
ferromagnetic regions of single domain dimensions controlled essentially by shape anisotropy. Magnetization changes occur by incoherent
rotation processes yielding coercive forces as high as 700 Oe. This is the
case for Cunico (see Table 6.2) which, like Cunife, produces permanent magnet behaviour when a ferromagnetic phase is precipitated.
The size and shape of the precipitated phase depends on the preparation
conditions: quenching from a high temperature plus cold working and
heat treatments can be optimized with regard to the resultant magnetic
hardness. Both phases have a face-centred cubic crystal structure in
Cunife, and Cunico. Anisotropic Vicalloy is also produced by a quench
cold-work treatment yielding very high values of remanent magnetization. Such materials are of best value in recording where high outputs
are required and low information storage densities are suitable.

Iron-nickel-chromium alloys are normally non-magnetic with a face-
centred cubic structure. However, on severe cold working, as for instance by drawing into a wire (HOBSON et al. [1948]), a ferromagnetic
body-centred cubic phase may be precipitated with a preferred orientation along the wire axis. The magnetic phase precipitates in acicular
form (typically $3 \times 0.1 \mu$) and is considered to be single domained.
The remanent magnetizations and coercive forces for these wires are
similar to the values obtained for oxide tapes (see Table 6.2); however,
their stability would be poor at elevated temperatures due to the low
Curie point (100° C).

Thin films of alloys, such as electrodeposited cobalt-nickel, have
been used as the magnetic recording medium for metal tapes and discs.
For these alloys a hexagonal structure is obtained for more than 72%
by weight of cobalt, and large values of saturation magnetization can
be achieved in the deposited layer (Table 6.2). It is found that a coercive force of approximately 250 Oe is obtained in films of a few microns

thickness. The ratio of remanent magnetization to coercive force is about 3, which is rather high if self demagnetization losses are to be avoided.

The coercivity of cobalt-nickel alloys may be considerably increased if the electrodeposition conditions are adjusted so that lamellated structured deposit is formed (SALLO and OLSEN [1961]). Considerable

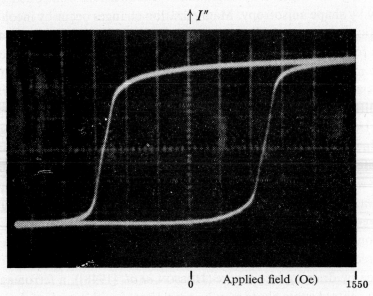

Fig. 6.18.   Magnetization loop for electrodeposited Co–Ni–P tape. $H_c = 800$ Oe. (KORETZKY [1963b])

shape anisotropy is thereby introduced and the resulting coercive force can exceed 1000 Oe. However, this effect is obviously achieved at the expense of reducing the saturation magnetization. Sufficient porosity is required to achieve single domain lamellae magnetically isolated from their neighbours. The method used is to combine electroless and electrolytic deposition (BONN and WENDELL [1953]). Sodium hypophosphite produces a reduction of the metal salts causing the electroless deposition, the most uniform results being obtained at elevated temperatures (about 50° C). Typical magnetic properties for this alloy are listed in Table 6.2 under cobalt-nickel-phosphorus, and it is seen that the coercive force and saturation magnetization are increased by a factor of 3 and 5, respectively, when compared to oxide powder tapes. Thus, for the same tape flux (or long wavelength output) the coating thickness may be reduced by a factor of 5, say to 2 microns. Since the short wavelength response depends on the intensity of magnetization rather than the total tape flux an increase of five times would then be obtained compared to iron oxide tapes. A 60 c/s hysteresis loop for a thinner layer is shown in Fig. 6.18; the high relative remanent magnetization is clearly indicated. Somewhat reduced magnetization and coercive force values are obtained on plating cobalt-phosphorus in the same manner (SALLO and CARR [1962]). The plating may be made onto metal bases as described or onto a plastic base previously coated with a conducting substrate. Cobalt has also been plated on mylar by an electroless process in which reduction is achieved in the presence of hypophosphite ions (FISHER and CHILTON [1962]). Typical samples have a saturation magnetization of 8 000 gauss and coercivity of 360 Oe. Such tapes, with a coating thickness of around 1.0 micron, show similar NRZ pulse packing and pulse width as iron oxide computer tapes, together with a slightly lower output.

Electroplating of other permanent magnet alloys has also been suggested in which multilayer deposits are made of materials with different magnetic characteristics (MATHES [1954]). In this way a high remanence, medium coercivity material (e.g. Cunife) may be plated with a medium remanence high coercivity material such as Cunico. As will be discussed in the next section, higher efficiency coatings may be produced if the coercivity is graded through the tape thickness.

Permanent magnet behaviour has been obtained in very thin layers of electrolytically deposited iron without the lamella structure (BOZORTH [1956]). However, it is necessary to reduce the thickness of the iron considerably to achieve high coercivity. Interleaving with a non-magnetic conductor would be a possible method to build a sufficiently thick layer for recording purposes. In this case the 100 % volume packing factor assumed for the saturation magnetization value in Table 6.2 is too high, and perhaps 50 % of this is nearer the practical value. Less success has been obtained in attempts to evaporate thin permanently magnetic metal layers, and the properties of the bulk material are obtained for small thicknesses. Evaporated cobalt layers, oriented by a magnetic field during evaporation yield a coercivity of 40 Oe for a thickness less than one micron (BLOIS [1955]). However, it can be expected that evaporation techniques will be developed in the future leading to controllable permanent magnet behaviour suitable for magnetic tape application.

§ 2. TAPE CONSTRUCTION

At present, the major factor distinguishing one tape from another is the perfection of its construction. Most manufacturers use the same plastic base and magnetic oxide materials and differ only in the binders and adhesives used to hold the magnetic layer. However, the recording properties of a tape depend substantially on the uniformity of the coating dispersion and the orientation of the particles in the magnetic layer. The smoothness of the coating surface and its ability to conform to the recording or reproducing head determine the uniformity of the recording and reproducing process. This is of particular importance in high density recording where the resolution is primarily determined by the intimacy of the head to tape contact. Although these mechanical considerations will always be of great importance for existing magnetic recording techniques it has been stressed that magnetic material developments, as outlined in this chapter, can be expected to provide major improvements in future magnetic tapes. Although the mechanical aspects of tape and tape transport design will not be treated in any detail, a short review of tape constructions is included here to illustrate the possible behaviour of the magnetic material in the recording and reproducing processes.

§ 2.1. *Plastic Base Material*

Three different types of base material have found commercial application in magnetic tapes. Cellulose diacetate has been used for many years; cellulose triacetate, has also been introduced more recently. Polyvinyl chloride, with room temperature properties somewhat

similar to triacetate (WERNER and KOHLER [1960]), has also found
wide application as a tape base material in Europe. A later development, polyester film, is now used universally for high quality tapes. It
has excellent physical characteristics, particularly for applications not
involving high temperatures, and its only major disadvantage is one of
cost.

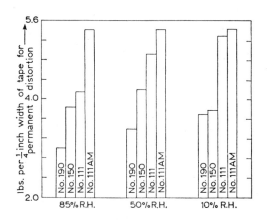

Fig. 6.19.   Thickness and humidity dependence of base distortion for acetate and
              polyester (PERSOON [1957]).
     Key: No. 190   – 100 gauge acetate
          No. 150   – 100 gauge polyester
          No. 111   – 150 gauge acetate
          No. 111 AM – 150 gauge polyester.

Cellulose diacetate and triacetate bases are produced by casting a
mixture of the acetate and a suitable plasticiser and solvent. Smooth
surfaces are produced on the cast films using this technique, allowing
corresponding uniformity of the magnetic layer coated thereon. A
disadvantage of the diacetate is its rather high humidity coefficient of
expansion, which can lead to buckling and curling (mechanical, not
magnetic!) of the tape, and consequent separation losses when run
over the recording or reproducing head. The humidity sensitivity of
triacetate is about half that of diacetate (SCHMIDT [1960]) and it has a
somewhat greater strength and flexibility. The latter characteristic is
particularly important for high density recording, and the tendency is

to make bases as thin as possible in order to increase flexibility and improve the conformability of the tape to the heads. A flexible tape will minimize the area of tape separated from the head when foreign particles carried on the tape coating cause a local separation. This factor is important with regard to the uniformity of the tape response. Of course, tape strength and dimensional stability requirements set a lower limit to the base thickness. Fig. 6.19 shows the thickness and humidity dependence of the force required to produce permanent distortion of the tape for acetate and polyester bases (PERSOON [1957]). It can be seen that reduction of thickness and increase in humidity reduce this critical force considerably and thus lead to greater difficulty in handling such a tape.

Polyester film enjoys almost complete immunity from humidity variations as can be seen from Fig. 6.19. This material is a single chemical compound, from which film is produced by extension and prestretching in two orthogonal directions. It is inherently flexible and does not require plasticisers. The prestretching processes produce orientation of the molecules leading to a strong stable product at room temperature. Irreversible shrinking occurs at high temperatures (about 5 % at 150° C), but such a process may be used to stabilize the material. The force required to produce 5 % elongation in polyester film is some 30 % higher than for equivalent thicknesses of acetate and its tear resistance is twice as high. A possible disadvantage of polyester film is its large elongation before breaking (100 %, compared to 15 % for acetate). Unidirectional pretensilized polyester can, however, greatly reduce this elongation before break in the tensilized direction. Thus, apart from high cost, polyester film can be considered to be an almost ideal base material for most magnetic tape applications. For extreme environmental conditions, new materials will have to be developed (STANLEY [1960]).

§ 2.2. *Dispersions and Coatings*

The degree of dispersion of the magnetic material in the plastic binder is a most important factor in determining the ultimate performance of the magnetic layer. Complete magnetic isolation of indi-

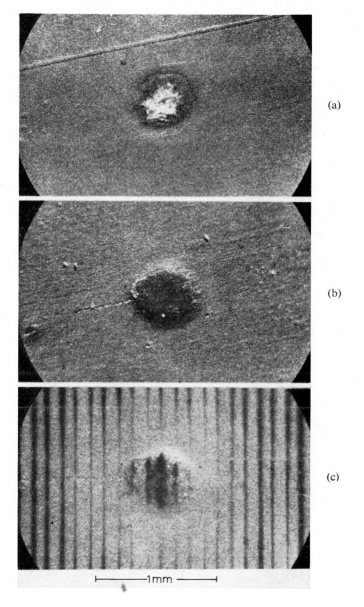

(a)

(b)

(c)

Fig. 6.21. Coating nodule and recorded signal pattern around fault in plastic base:
a. Blemish in plastic base. b. Coating nodule due to base fault.
c. Surface field pattern of recording near nodule.

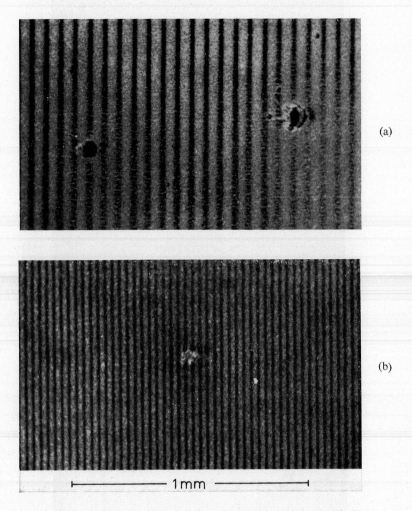

Fig. 6.22. Surface magnetization patterns around small coating faults.
a. nodules;   b. hole.

vidual particles has not been demonstrated in dilute dispersions, where
the magnetic properties suggest that some agglomeration persists even
after extensive ball-milling (see § 1.1). Nevertheless, milling techniques
have been optimized by tape manufacturers, yielding sufficiently good
dispersions to allow a considerable degree of particle orientation to
take place in a magnetic field. In addition, the dispersions achieved
lead to low enough background noise levels in tape that a change of
particle size in a dispersion is manifest as a change in these levels.

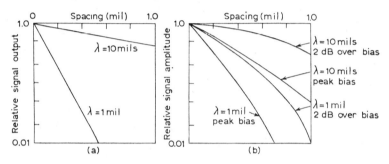

Fig. 6.20. Spacing losses in (a) reproducing, and (b) recording
(VON BEHREN [1958]).

Several different techniques are used in the tape industry to coat the
dispersion onto the plastic base. In a common simple method, called
knife coating, the plastic base is transported under a hopper containing
the dispersion. The trailing edge of the open underside of the hopper is
smooth and flat and is held parallel to the upper surface of the moving
tape at a distance of several times the required dried coating thickness.
All else being equal, the coating thickness is constant for constant base
thickness, and the coating surface smoothness is determined by the
perfection of the coating blade. In another coating technique, known
as the gravure method, a highly polished coating roll running in contact
with the plastic base is continuously coated with dispersion from a feed
roll. In this way a constant thickness of dispersion is laid onto the
coating roll and transferred to the tape. For this method the coating
surface is as smooth as the coating roll. Both techniques can yield a

uniform coating with a smooth surface. The latter property is of particular importance when short wavelengths are to be recorded since both recording and reproducing losses are very high when the effective head to tape separation is of the order of the recorded wavelength. Figs. 6.20*a* and *b* show the reproducing and recording losses, respectively, indicating that the recording loss may be somewhat reduced by increasing the amplitude of the ac bias (VON BEHREN [1958]); however, for high resolution recording, the longitudinal decrement of the recording field must be high and this condition only occurs close to the recording gap.

Local separation of the coating from the recording head can occur due to faults in the tape surface, or foreign particles trapped between the head and the tape, or creases in the tape. In most cases, the area of tape effectively separated from the heads is greater than the area of the fault itself. This is illustrated in Figs. 6.21 and 6.22 which show recordings made visible with colloidal magnetite (see e.g. MEE [1958], ELMORE [1938]). The large blemish in the plastic base shown in Fig. 6.21*a* causes a coating nodule which lifts an area of about twice its own diameter sufficiently away from the recording gap to prevent signal recording. Smaller coating faults are shown in Fig. 6.22 in which loss of resolution over a wide area is shown to occur for nodules (Fig. 6.22*a*), whereas a similar size hole in the coating (Fig. 6.22*b*) causes no such separation loss. Some improvement of the surface smoothness of tape coatings may be obtained by calendering, using highly polished rolls. Also, due to the reduction of coating thickness on rolling, a densification of the magnetic powder occurs leading to improved short wavelength response.

§ 2.3. *Orientation in Coating*

The general benefits to be gained from orientation of uniaxial anisotropy particles have been described earlier in terms of the increased signal output and the reduced background noise level. Higher output is obtained at both long and short wavelengths when the direction of the recording field is essentially in the direction of orientation. The magnetic properties giving rise to this improvement are an increase in remanent magnetization, I_r'', before saturation effects are evident, and an increase in anhysteretic remanent magnetization in the linear region of the anhysteretic magnetization curve. Thus, the dispersions of the particle switching fields and of the internal fields appear to be reduced by orientation. It has been indicated in Chapter 2, § 6 that a restriction of the spread of effective switching fields would also lead to more efficient short wavelength recording, when the signal changes its amplitude during the ac bias decrement.

A drawback with orientation of uniaxially anisotropic particles is

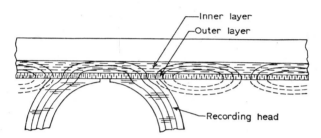

Fig. 6.23.   Two-layer tape with particle orientation in two directions (GABOR and BAUER [1962]).

that the magnetic and the recording properties are only substantially improved when the applied field is in the orientation direction. Thus, when a wide range of wavelengths is to be recorded, greatest efficiency would be obtained if the orientation direction conformed to the recording field direction at different depths in the coating. This would mean that the layers remote from the recording head should be oriented parallel to the direction of tape motion, whereas the surface layers should be oriented almost perpendicular to this direction. A two layer tape, with the inner layer containing longitudinally oriented particles and the outer layer perpendicularly oriented particles, is illustrated in Fig. 6.23 as a possible practical method for implementing such a tape coating (GABOR and BAUER [1962]). Evidence that the recording field direction is at an angle to the longitudinal direction may be obtained by measurements at short wavelengths where only the surface layer of the tape is active. When the orientation direction is tilted away from the longitudinal direction a difference of tape sensitivity of up to 6 dB may be obtained when transporting such a tape in opposite directions (VON BEHREN [1955]). A further limitation occurs for applications requiring tape to be magnetized in different directions. Examples of such applications are found in magnetic disks and the transverse track type of video recording combined with longitudinal track sound recording.

Orientation may be produced in the coated tape immediately after the dispersion has been spread on to the base material. The viscosity of the coated layer, degree of dispersion and dilution, as well as the acicularity of the particles, will all have some influence on the amount of orientation produced. Some orientation occurs due to the longitudinal shearing effects during the coating process, but a magnetic field applied to the wet tape produces a substantial increase of orientation. The direction of this field is so adjusted that the particles settle down into a longitudinal orientation as they pass away from the field. It is found that orientation may be produced if the tape runs along the axis of a dc solenoid, or if it passes in the vicinity of a permanent magnet. In the latter case a sufficiently strong orienting field is obtained close to the magnet and the particles experience this field for a short time only. However, the orienting action may be looked upon as one of producing unidirectional magnetization followed by a relief of local torques by particle rotation. The particles with their easy axis at about 90° to the orienting field are the most difficult to orientate and yet are the most important in producing a higher remanent magnetization. It can be assumed that for such particles coherent rotation of magnetization will occur and eq. (5.7) is valid. That is to say, for an angle θ between the magnetization and the easy axis the aligning couple (C) is given by

$$C = dE_i(\theta)/d\theta$$
$$= 0.5 \, vI_s^2(N_b - N_a) \sin 2\theta. \tag{6.10}$$

This is a maximum when $\theta = 45°$, and zero when $\theta = 0$ and 90°. Thus, the most effective aligning field is that which turns the magnetization through 45°. For particles at 90° to the applied field the equilibrium condition between anisotropy field and applied field energy density is

$$I_s^2(N_b - N_a) \sin \theta \cos \theta - H I_s \cos \theta = 0 , \qquad (6.11)$$

whence, for $\theta = 45°$,

$$H_{crit} = I_s(N_b - N_a)/2^{\frac{1}{2}} . \qquad (6.12)$$

For γ Fe$_2$O$_3$, and a particle length/width $= 6$, H_{crit} is 1500 Oe. Higher fields will produce more orientation from this position but it is necessary for the orienting field to pass through the above critical value to achieve maximum orienting torque on particles at 90° to the applied field direction.

§ 3. MAGNETIC MEASUREMENTS

In this section techniques are described for measuring some of the magnetic properties used to evaluate magnetic materials for tape and some of the recording properties of new experimental sample tapes. Many other measurements of tape properties, both physical and magnetic, are performed on commercial tapes and these have been adequately reviewed elsewhere (SPRATT [1958]).

§ 3.1. *Measurement of Magnetization Characteristics*

Dynamic Methods

The most common techniques for measurement of magnetization of tape materials are those involving a change of magnetization level in the sample, and detection of the change by means of a search coil. Open magnetic circuit measurements are usually satisfactory for magnetic tape since the saturation magnetization is low and demagnetizing factors are small. Powder samples can be packed into slender tubes, necessitating only moderate sample lengths to obtain small demagnetization losses. Should short samples be used, requiring compensation for the demagnetizing field due to the free poles on the sample, a useful technique is to provide compensating coils on the ends of the sample which produce an equal and opposite field to the demagnetizing field. Compensation is complete when the field in the vicinity of the sample is equal to the applied field with the sample absent.

Three types of sample excitation are in common use. They may be classified as switched magnetizing field reversals, continuous field cycling, and pulsed excitation fields. The first method, the so-called ballistic testing method, is the classical technique for measurement of magnetization properties of magnetic materials. Hysteresis loops may be plotted by measuring the flux changes, occurring for step-by-step magnetization changes, using a search coil and ballistic galvanometer. This technique is gradually being superseded owing to the tedious and lengthy measurement procedure and, in some cases, due to the uncertainty of time effects in magnetization changes which may affect the measured changes when sudden changes occur in magnetizing field levels. This latter problem has also to be considered when using pulsed magnetizing fields. Continuous field changes are more commonly used in magnetization loop measurement of tape materials.

Depending on the type of magnetic material to be tested and its prospective use, continuous loop cycling equipment may range from a rate of many seconds per cycle to frequencies in the kilocycle per second region. The slowest cycling speeds are used in testing high permeability samples where large flux changes occur in a small fraction of the cycle time. Many dc hysteresigraphs have been designed for this purpose, using, for instance, manual control of the magnetizing current and integrating galvanometer amplifiers for measurement of the sample induction (CIOFFI [1950], BOCKEMUEHL and WOOD [1960]).

If an air-cored magnetizing coil is used with this type of instrument, the applied field may be conveniently calculated from the magnetizing coil current. If, on the other hand, the sample has a high coercive force, then an air-cored coil may not be adequate to saturate it. In this case the required magnetizing field may be produced by an electromagnet. Due to a possible variation of the applied field along the sample it is preferable to measure the applied field directly in the vicinity of the pickup coil. This may be achieved with a flat search coil alongside the sample, or with a magnetic potentiometer as illustrated in Fig. 6.24. An integrating galvanometer amplifier then yields a measurement of the applied field (BROCKMAN and STENECK [1954]). The intrinsic induction ($B_i = B-H$) may be measured by subtraction of the appropriate proportion of the field coil output from the flux coil output as shown in the inset of Fig. 6.24. The outputs of the integrating amplifiers are fed to an X-Y recorder which plots the B_i vs. H curves directly.

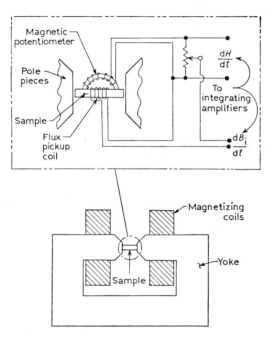

Fig. 6.24.   Hysteresis curve measurement using electromagnet for field generation.

A simpler method for continuous magnetization loop plotting is obtained when the sample excitation is obtained from an ac power supply. Air-cored magnetizing coils may be conveniently resonated at the power frequency. The intrinsic induction of the sample is measured by means of a search coil wapped closely around the sample: the component of the induced voltage due to the solenoid being bucked out by a series-opposition connected coil having the same flux linkages with the magnetizing coil. Adequate integration may be achieved using a resistance–capacitance network compensated for phase shift

(WIEGAND and HANSEN [1947]). The peak applied field may be measured via a series resistance in the solenoid circuit, or by integrating the secondary voltage of a known mutual inductance whose primary is in the solenoid circuit. The field and magnetization voltages so produced may be applied to the x and y plates of an oscilloscope on

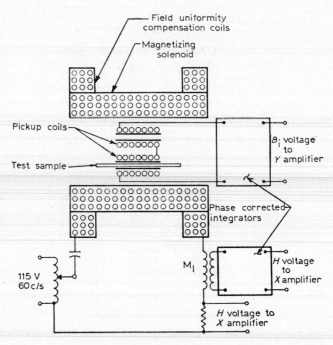

Fig. 6.25.   Power frequency hysteresis loop measurement apparatus.

which the sample hysteresis loops will be displayed. A schematic diagram of this type of equipment is shown in Fig. 6.25.

If a vertical deflection d on the oscilloscope is obtained from a sample of length 1 cm, weight m g, and cross sectional area a cm^2, the tape magnetization (I'') is given by

$$I'' = f d / 4\pi a, \tag{6.13}$$

where f is the flux sensitivity factor. If the sample is porous, the intensity of magnetization (I) of the material in the sample is then

$$I = I''\rho'/\rho,\tag{6.14}$$

$$\text{i.e.}\quad I = f\,dl\rho'/4\pi m,\tag{6.15}$$

where ρ' and ρ are the material and sample densities respectively.

Although the calibration for flux sensitivity may be made using a standard sample, a better method is to introduce a measurable flux into the pickup circuit. This may be achieved by placing a single layer solenoid of accurately known dimensions into the pickup coil, or by producing a known field from the magnetizing solenoid; the latter may be calculated accurately from the solenoid dimensions.

In certain circumstances, if the B and H waveforms contain odd harmonics only, B_i and H may be measured by rectifier type galvanometers connected directly to the sample flux coil and the secondary of the mutual inductance. A more versatile technique (EDMUNDSON [1955]) uses a mechanical relay to control the galvanometer current. The relay acts as a rectifier allowing the galvanometer to read the average output during the time of closure of the contacts.

Thus, by connection to the pickup coils or to the secondary of the field measuring mutual inductance, the average galvanometer current (i_{av}) is respectively given by

$$i_{av} = \frac{nf}{R}\int_{t_2}^{t_1}\frac{d\varphi''}{dt}\cdot dt = \frac{nf}{R}(\varphi''_{11} - \varphi''_{12}),\tag{6.16}$$

or

$$i_{av} = \frac{Mf}{R}\int_{t_2}^{t_1}\frac{di}{dt}\cdot dt = \frac{Mf}{R}(i_1 - i_2),\tag{6.17}$$

where n = number of turns on sample pickup coil

f = magnetization cycling frequency

R = galvanometer circuit resistance

M = mutual inductance

i = magnetization current

$\varphi''_i = B''_i a$ = sample intrinsic flux

If the time $(t_1 - t_2)$ for which the relay is closed is half a cycle then $\varphi''_{i11} = -\varphi''_{i12}$, $i_1 = -i_2$, and $H_1 = -H_2$ as shown in Fig. 6.26. By switching the relay controlled galvanometer to the H and φ''_i coils corresponding values of H and φ''_i are obtained. Variation of the switching phase enables complete hysteresis loop data to be obtained. By shortening the time for which the relay is closed the loop may be explored in detail. A modification of this type of instrument employs semiconductor

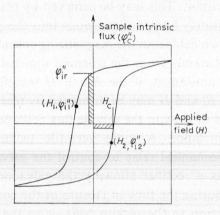

Fig. 6.26.   Diagram of switched measurement points on hysteresis loop.

diodes, instead of the relay, which are controlled in their switching phase with respect to the magnetizing current (GEYGER [1956]).

The remanent intrinsic induction and the coercive force, may also be measured by appropriate electronic switching of peak reading voltmeters to the B_i and H outputs of the electronic integrators described earlier. The switched voltmeters may be triggered to read B_{ir} and H_c when H and B_i pass through zero, respectively, as illustrated in Fig. 6.26 (KRAMER [1956]).

The use of pulsed magnetic fields allows the production of very large magnetic fields for a short period of time. One method employs a capacitor discharge into the magnetizing coil producing a damped oscillation of magnetizing field (JACOBS and LAWRENCE [1958], ELDER and BARTE [1962]). Field and intrinsic induction may be measured using a relay controlled ballistic galvanometer in which the phase and time interval of the galvanometer integration can be controlled. On comparing hysteresis loops for γ Fe$_2$O$_3$ powder produced by the pulse technique and the classical field reversal method, it is observed that similar results are obtained. Eddy current effects are presumed to be negligible in this material. However, the pulse field measurements do indicate that H_c increases slightly with H_{max} when the latter is greater than the value required to accomplish complete irreversible magnetization of the sample. This has been attributed to magnetic viscosity effects causing a lag of the magnetization behind the applied field. It is further shown that $H_{max} > 50 H_c$ is required to produce 99% of saturation magnetization in γ Fe$_2$O$_3$ (ELDER and BARTE [1962]).

Magnetometer Measurement of Magnetization Characteristics

Magnetometer methods are used in tape materials research to measure the magnetic moment of small samples of powders and its variation with temperature. Certain types of magnetometers are also well suited for measurements on the angular variations of magnetic properties of anisotropic samples. The magnetization of tape recorded with a periodic variation of magnetization level may also be measured by a magnetometer technique which determines the external demagnetizing field of the recording. Since magnetic materials for tape usually have low saturation magnetizations and relatively high coercive forces, self demagnetization losses are low and corrections for samples with large demagnetization coefficients are reasonably accurate.

Although there are several forms of magnetometers in use, greatest versatility and simplicity is obtained in vibrating sample magnetometers, particularly if the direction of vibration is orthogonal to the applied field direction. This technique was first used in the classical magnetization reversal type of measurement by WEISS and FORRER [1929] for spherical samples. The sample was withdrawn from the field of an electromagnet in a direction perpendicular to the field. The change of field near the sample, due to its movement, was detected with coils in Helmholtz arrangement placed between the sample and the exciting magnet poles.

Much greater sensitivity and ease of measurement have been obtained by using periodic motion of the sample, for instance by vibrating the sample in the vicinity of the search coil. A high input impedance ac voltmeter may be used to measure the pickup coil voltage, allowing a large increase in the number of search coil turns compared to the ballistic method. If sample magnetization is obtained from a solenoid

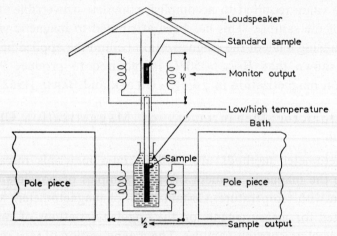

Fig. 6.27.   Vibrating sample magnetometer.

then the simplest technique is to vibrate the sample along the coil axis parallel to the field (VAN OOSTERHOUT [1956]). Vibration parallel to the field may also be accomplished using an electromagnet to supply the field (FLANDERS [1957]), but it is then necessary to modify the magnet considerably to achieve the necessary motion. In the method described here, vibration perpendicular to the applied field direction is simpler and angular variations of magnetic properties may be easily measured (FONER [1959]). The apparatus layout for this method is shown in Fig. 6.27. A frequency of vibration of about 100 c/s with an amplitude of about 1 mm is convenient and may be obtained either with a loud-speaker, as shown, or by means of a synchronous motor with eccentric drive. The amplitude and frequency of vibration may be checked by means of the monitor coils and standard sample illustrated. By distinction from the previous techniques, vibration of a magnetized sample perpendicular to the field is detected by search coils (output V_2) not coaxial with the vibration axis. The double search coil configuration shown in Fig. 6.27 is particularly suitable since small variations in the sample position have little effect on the sensitivity. In addition, the double coil is insensitive to external fields and to instabilities in the electromagnet field.

For measurements on magnetic powders a suitable form for the sample is a thin cylindrical sample of packed powder as shown in Fig. 6.27. The cylinder length is sufficient to avoid end effects and thus makes the vertical position of the sample non-critical. The magnetometer may be calibrated with a sample of known magnetic moment, such as high purity nickel, having similar dimensions to the powder cylinders. Alternatively, absolute calibration can be obtained by vibrating a current carrying solenoid having the same dimensions as the sample. Using a high permeability standard sample, such as nickel, in a form with a large known demagnetizing coefficient (e.g. cylinder or sphere) the magnetization in small fields is governed entirely by the demagnetizing field, and the internal field is zero. Hence, calibration may be made in terms of the known demagnetizing coefficient rather than the known saturation magnetization. The magnetization sensitivity factor K is then given by

$$e_{\text{cal}} = KvI \,, \qquad (6.18)$$

where e_{cal} is the pickup coil voltage, and v the standard sample volume. For a spherical sample, the applied field $H_{\text{appl}} = 4\pi I/3$, and for a cylinder, $H_{\text{appl}} = 2\pi I/3$. Thus, for the cylindrical standard sample,

$$e_{\text{cal}} = 3KvH_{\text{appl}}/2\pi \,. \qquad (6.19)$$

The intensity of magnetization of the powder in a cylindrical sample is then given by

$$\begin{aligned} I &= (e/Kv') \, (\rho'/\rho) \\ &= e\rho'/KM \,, \end{aligned} \qquad (6.20)$$

where ρ is the sample density $= \dfrac{\text{sample mass}}{\text{sample volume}} = \dfrac{M}{v'} \,,$

and ρ' the material density.

The specific magnetization (σ) is then given by

$$\sigma = e/Kv'\rho . \tag{6.21}$$

From eq. (6.20) it is seen that the true density of the material must also be measured in order to determine the intensity of magnetization of the sample. Using the method of displacing a liquid leads to errors with very fine powders unless care is taken to ensure complete wetting of the sample. A differential measurement using displacement of a gas gives rapid and accurate results.

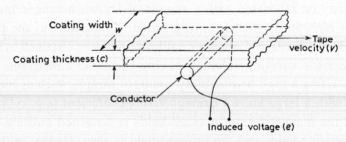

Fig. 6.28. Measurement of tape magnetization with circular cross-section conductor.

Measurement of Recorded Tape Magnetization

A convenient alternative method for measuring the magnetization level in a tape is to record a sinusoidal flux pattern on the tape, and to detect the surface demagnetizing field by transporting the tape over a single conducting wire, as shown in Fig. 6.28. A small voltage e, proportional to the normal induction at the surface of the tape B''_y, is induced in the conductor. For a circular conductor of radius r, it has been shown (DANIEL and LEVINE [1960b]) that the peak voltage $\hat{e}$ is related to the peak normal induction $\hat{B}''_y$ by the relation

$$\hat{e} = vw\hat{B}''_y \exp\left(-2\pi r/\lambda\right) , \tag{6.22}$$

which indicates that the wire acts like a filament of infinitesimal cross-section at a distance r from the tape. For peak level recording, on a conventional oriented acicular iron oxide powder tape, the rms value of the surface induction is of the order of 40 gauss, and about 10 gauss for normal recording levels.

Assuming the recorded magnetization, I''_{ar} (H_{sig}), to be uniform through the tape, as is the case for $\lambda \gg c$, the relationship between the normal induction and the tape flux (φ'') is given by

$$w\, B''_y = 0.5\{\, \mathrm{d}\, [\hat{\varphi}''\cos(2\pi x/\lambda)]/\mathrm{d}x\}$$ (6.23)

whence

$$\hat{B}''_y = 4\pi^2 \hat{I}''_{ar}\,(H_{sig})\, c/\lambda .$$ (6.24)

It is assumed here that the tape magnetization is entirely longitudinal, a condition which is satisfied providing the ac bias level used to produce the recording is not greater than that giving peak tape magnetization $(\hat{I}''_{ar}(H_{sig}))$.

Good agreement has been obtained between tape magnetization measured with a single conductor, and that measured on a sample magnetized with an equivalent dc signal; remanent flux measurement for the latter case is achieved by cutting the tape into strips several inches long and withdrawing a bundle of such strips from a coil connected to an integrating galvanometer. The main difficulty of the single conductor technique is in the measurement of the extremely small voltage developed. Greater sensitivity would be obtained by using a reproducing head which may be calibrated using a standard tape recorded with a signal of known flux level; however, the long term accuracy would be poor due to changes in the reproducing head with use.

§ 3.2. *Measurement of Anhysteretic Magnetization*

In addition to the remanent magnetization acquired from dc applied fields it is necessary to know how this parameter depends on a combination of dc and ac field conditions. Magnetization may be acquired anhysteretically by various combined cycles of ac and dc fields. Those of prime interest in recording are the ideal process, in which the ac field gradually reduces from a saturating value to zero while the dc field remains constant, and the modified process in which the ratio of ac and dc is kept constant during the field reduction to zero. Due to the complexity of these magnetization programmes it is not convenient to use a periodic cycling method of measurement and so ac flux detection schemes are inapplicable.

Anhysteretic remanent magnetization may be achieved with the equipment shown in Fig. 6.29, in which the applied ac field may be

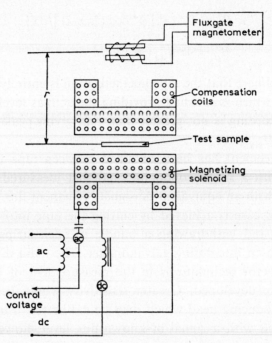

Fig. 6.29.   Apparatus for anhysteretic magnetization measurement.

slowly reduced to zero for constant values of simultaneously applied dc field. To obtain true anhysteretic magnetization conditions it is necessary that the reduction in ac amplitude per cycle be small compared to the dc field, a condition which is easily achieved if the ac is reduced to zero in about 20 seconds. The modified anhysteretic magnetization conditions may be obtained by setting the ac and dc levels and slowly withdrawing the sample from the coil; alternatively, the same effect is obtained by controlling the dc supply with the ac Variac as shown in Fig. 6.29.

With both types of measurement it is necessary to use a sample with a small external demagnetizing coefficient to avoid shearing the measured curves. The effect of a demagnetizing field on anhysteretic magnetization is considered in Chapter 2, § 4, where it is shown that the product of the demagnetizing coefficient and true anhysteretic susceptibility (s) should be about 0.01 for a 1% error in apparent susceptibility (eq. 2.9). This condition is not hard to meet in oxide tape samples where s is about 4, and a sample a few inches in length, consisting of several strips, is satisfactory. For samples of metal powders and alloys the anhysteretic susceptibility may be very large and it is more convenient to close the magnetic circuit with an external yoke to reduce the demagnetizing field (see e.g. GOULD and McCAIG [1954]).

The remanent magnetization of the tape sample can be measured in a number of ways. A second harmonic fluxgate type magnetometer can be used to give a continuous reading of the demagnetizing field, H_d, of the sample. If it is placed in the Gauss $'B'$ position, shown in Fig. 6.29, the demagnetizing field is related to the remanent magnetization, I''_{ar} by,

$$H_d = I''_{ar} v / r^3 , \qquad (6.25)$$

where r is the magnetometer–sample distance, which is large compared to the sample length. Alternatively, the magnetization may be measured by vibrating the sample as described previously, or a ballistic type of measurement may be made by integrating the voltage developed when the sample is withdrawn from a search coil. Using a time constant, RC, for integration, the maximum voltage, $\hat{e}$, developed across the capacitor on withdrawing the sample from a search coil of n turns is given by

$$\hat{e} = naI''_{ar} 10^{-8} / RC , \qquad (6.26)$$

where a is the cross sectional area of the sample. Very long time constants and high sensitivity may be achieved with an electronic integrating amplifier (SCHMIDBAUER [1957]); calibration of the flux sensitivity is obtained by creating a known flux change, $\Delta\varphi_{cal}$, in the secondary of a mutual inductance M whence

$$\Delta\varphi_{cal} = M\Delta i / n_1 , \qquad (6.27)$$

where Δi is the primary current change, and n_1 the secondary turns.

§ 3.3. *Measurement of Tape Recording Properties*

In this chapter the properties of magnetic materials for tape have been described and discussed in terms of their fundamental magnetization processes. Performance criteria for magnetic tapes were established in terms of the magnetization characteristics of the magnetic carrier and methods for measuring these characteristics have also been described. It remains, then, to consider how the various materials perform when recordings are made on the corresponding tapes. The recording performance of tape is also dependent on the physical qualities of the magnetic layer and the plastic carrier. In this discussion of tape characteristics which are directly dependent on the magnetic material properties, it is assumed that equivalent physical quality has been obtained in the various samples discussed.

The basic tape characteristics through which the recording performance of a material may be assessed are,

(i)   Output capability.
(ii)   Recording sensitivity.
(iii) Frequency response.
(iv)   Tape noise.
(v)   Layer to layer print-through.

The output capability of tape can be measured in an absolute manner by determining the maximum normal surface induction, for a specified low distortion sinusoidal recording, using a single conducting wire as previously described. As concluded in Chapter 2, the ac bias recording process is essentially a modified anhysteretic magnetization process in which different layers of the coating experience combined signal and bias fields of constant ratio, but different maximum values. In agreement with the theoretical analysis of the anhysteretic magnetization process, most tape materials have similar remanent anhysteretic magnetization characteristics which differ only in the maximum value. The characteristics are markedly dependent on the maximum ac field, however (see Fig. 2.13, Ch. 2) and the average characteristic obtained in recording is more linear than that obtained for a constant maximum field applied to all elements in the tape.

Consequently, the output capability for a specified distortion will depend on the field gradient through the coating. It is preferable, then, to use a constant ratio of coating thickness to recording head gap length when determining the long wavelength output capability.

The recording sensitivity of a tape is also dependent on the decrement of the recording field through the thickness of the magnetic layer, and it is usual to specify this in terms of the recording head gap length and tape thickness used for the test. Since, for ac bias recording, the elements of the tape are subjected to a modified anhysteretic magnetization process it should be possible to predict the recording sensitivity, knowing the anhysteretic susceptibility characteristics and the field gradient through the coating. However, as described in Chapter 2, § 6, other effects appear to take place in recording to make an accurate prediction difficult. From the point of view of measurement of an effective tape sensitivity it is convenient to use a standard recording head gap length, such that the field gradient is not large and is held constant for subsequent tests. On recording and reproducing with this head, using an ac bias level giving maximum sensitivity, the recording current i_s and reproducing voltage e may be measured for a low frequency signal. These are then used in the following simple analysis to yield a measurement of effective tape sensitivity (DANIEL and AXON [1955]).

Assuming linearity, and constant applied field, H_x, through the tape thickness, the tape sensitivity η_t'' is given by

$$I_{ar}''(H_{sig}) = \eta_t'' K i_s = \eta_t'' H_x, \qquad (6.28)$$

where i_s is the recording head current, and K a constant.

Further, assuming reciprocity, the reproducing head voltage e, produced on playing back the recording made by i_s, is

$$e = K' \eta_t'' i_s / \lambda, \qquad (6.29)$$

where

$$K' = 2\pi v w c g K^2,$$

and v is the tape velocity, w the tape width, c the tape thickness, and g the recording head gap length.

Combining eqs. (6.24), (6.28) and (6.29) gives an expression for the effective tape sensitivity in terms of measurable quantities,

$$\eta_t'' = v\lambda wg(\hat{B}_y'')^2/8\pi^3 cei_s .\tag{6.30}$$

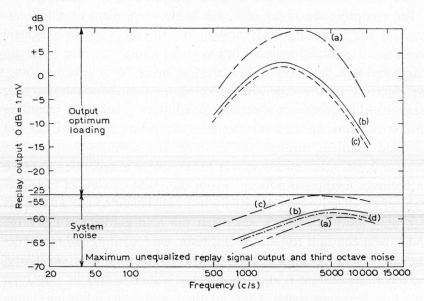

Fig. 6.30.   Recording performance of powders:
a. Acicular alloy powder 0.015μ.   c. Acicular particles 1.0μ.
b. Acicular particles 0.2μ.   d. Non-acicular oxide powder 0.1μ.
Speed $1\frac{7}{8}$ inch/sec.

The maximum surface induction $(\hat{B}_y'')$ may be measured using a single loop for reproduction, as already described.

The most important characteristic distinguishing one tape from another is its short wavelength response. However, the wavelength dependence of the recorded flux is extremely difficult to measure accurately owing to the uncertainty of the losses in the reproducing process. The single turn loop head is not useful here owing to its limited resolution, determined by the thickness of the wire. Reproducing heads have wavelength dependent losses due to the finite length of the gap and the imperfect contact between the tape and the head (see Ch. 4, § 2); frequency dependent losses also occur due to eddy currents in the head core but these may be kept negligible by testing tapes at very slow speeds rather than at high frequencies. The technique of making ultrafine gap reproducing heads has now reached the stage where head sensitivities may be held within a few decibels for repro- duction of wavelengths down to about 4 microns. Thus, comparison of frequency response and output capability is possible if high preci- sion recording and reproducing heads are used.

The recording performances of some of the tape materials described in this chapter are compared in Fig. 6.30, where the maximum repro- ducing head output is shown as a function of frequency, using a slow tape speed of $1\frac{7}{8}$ inch/sec. Curves are drawn for a thin layer tape of alloy powder, and for thicker layers of oxide powder tapes having comparable remanent fluxes and hence comparable long wave-length responses. These curves include two acicular particle tapes having different particle sizes.

The curves correspond to a harmonic distortion level of 5 %, or less. As is described in Chapter 7, the output curves depend in a complicated way on the bias amplitude; this is chosen as a compromise between overbiassing the short wavelengths and underbiassing the long wave- lengths, and is a function of the field gradient through the thickness of the coating. A better compromise bias is achieved in thinner coatings. In the curves shown, the bias is optimized for the mid frequency range and it is clearly seen that the thin layer of high magnetization alloy powder is superior in overall output. This is due to minimizing the loss due to variations of bias level through the coating.

Tape noise and layer-to-layer print-through measurements are dealt with in detail in Chapter 4. However, the effect of particle size on the zero modulation noise may be seen in Fig. 6.30. Here the noise, measured in third octave bands is plotted for four tapes and again the superiority of the alloy powder, with its extremely small particle size, is evident.

The effects of the recording properties of tapes on the performance of various tape recording systems are discussed in the following chapter.

CHAPTER 7

EXPERIMENTAL RECORDING TECHNIQUES

§ 1. INTRODUCTION

The main body of this book is concerned with the basic processes whereby a magnetic tape may become magnetized in accordance with a time varying signal, and how this magnetization may be detected. External to the recording and reproducing transducers, and to the tape itself, the electrical signal to be recorded may be electronically tailored to be in a form most suitable for recording. Complementary tailoring of the reproducing head voltage is also, then, required to retrieve the original signal. It is necessary to consider all of the processes involved in recording, storage, and reproduction of an electrical signal as a complete system in order to produce an optimum result. The nature of the original signal to be recorded, be it analogue or digital in form, and the characteristics of different recording techniques, be they direct signal recording with bias or various modulation techniques prior to recording, all have a bearing on the overall system design.

When a large range of frequencies is to be recorded, direct recording encounters the problem of the inherent differentiation process occurring between the recording and reproduced signals. The consequent frequency dependent amplitude correction, known as equalization, can be very large and difficult to achieve. It is sometimes preferable to restrict the frequency band to be recorded by the use of some carrier system, thus avoiding concomitant long and short wavelength recording problems and large equalizations. Of course, the carrier frequency must be higher than the highest signal frequency, necessitating an increase in tape speeds and entailing a reduction of information packing density on the tape. Furthermore, new critical factors arise in the recording system. In the case of amplitude modulation of a carrier, the random amplitude modulation occurring in the recording-repro-ducing system must be minimized; this may be accomplished best by increasing the recorded wavelength through a further increase of tape speed. Frequency modulation, on the other hand, demands that changes of tape speed be held low enough so that the recording system's inherent frequency modulation (wow and flutter) is negligible. Fre-quency modulation has been successfully applied to wide-band video and instrumentation recording systems where the carrier is recorded without bias at maximum level at a short wavelength. Amplitude linearity is unimportant unless more than one FM carrier is to be re-corded on the same track, in which case some biassing system must be used.

Another method of electronically tailoring a signal to suit the system is to digitize it. Some signals are already in digital form and can be easily converted into a binary code suitable for NRZ or RZ saturation recording, as described in Chapter 3, § 1.2. Analogue signals may be sampled at regular intervals and each sample converted to a group of binary type pulses coded to represent the magnitude of the sample. A variety of techniques for data recording of pulses exists, employing the basic recording mechanisms described in Chapter 3; these techniques have been recently reviewed elsewhere (DAVIES [1961]).

Notwithstanding the ever-growing number of recording techniques, particularly in the data recording field, direct recording is by far the most efficient with regard to information packing density on the tape. Much of the understanding of the recording and reproducing proces-ses, and the recent advances in heads and tapes described in this book, are directly applicable to high-resolution wide-band recording systems. It is, therefore, appropriate to describe the design of an ac bias recording system in some detail to illustrate the application of im-proved magnetic recording components and to point the way for future developments. The present state of progress is such that audio record-ing, embracing the frequency range from 30 to 15 000 c/s, may be satisfactorily performed at a tape speed of $1\frac{7}{8}$ inch/sec. Higher packing densities have been achieved in the laboratory where 13 500 cycles per inch of tape have been reported for a wide-band system

(BROPHY [1960]), which would reduce the tape speed for the above system to about 1 inch per second.

Wide-band high-resolution recording systems encounter the problem of simultaneous efficient recording of long and short wavelengths. This problem may be alleviated to some extent by special recording head designs. Amplitude equalization can also be judiciously apportioned between the recording and reproducing channels to produce the maximum recorded signal level on the tape with acceptable distortion. The development of an optimum system taking account of these factors, in addition to the improvements in tapes and heads, will be discussed in this chapter.

§ 2. WIDE-BAND HIGH-RESOLUTION RECORDING SYSTEM

§ 2.1. *Design Considerations*

Since the system to be considered involves the recording and reproduction of wavelengths, both long and short compared with the thickness of the magnetic coating and the recording head gap length, it is necessary to take into account both the "long" and "short" wavelength recording functions described in Chapter 2, § 6 and the corresponding reproducing functions described in Chapter 4, § 2 and 4, § 5. However, before this is done, it is necessary to ascertain the subjective effects of distortion of the signal and of the noise of the system in order to establish boundary conditions for component design.

Odd harmonic distortion occurs in ac bias recording and it has been determined that this has negligible annoyance if the harmonic amplitude at the reproducing head coil is less than 5% of the total signal. This assumes that the transfer characteristic $\dfrac{\text{reproducing head voltage}}{\text{recording head current}}$ rises at 6dB per octave due to the inherent differentiation in the system. However, when the recorded frequency exceeds 5 kc/s the harmonic distortion lies outside the pass band of the system. Even at somewhat lower frequencies the distortion components become attenuated due to the reduced flux-carrying capacity of the tape at short wavelengths. This factor will be discussed further when equalization techniques are considered. Due to these losses relatively higher recording levels are tolerable. The limiting effect at high magnetization levels is then set by non-harmonic distortion components occurring in the audio frequency band; these are due to the intermodulation of signal frequencies recorded simultaneously in a system with a non-linear transfer characteristic. It is found that magnetization levels can approach the maximum record-

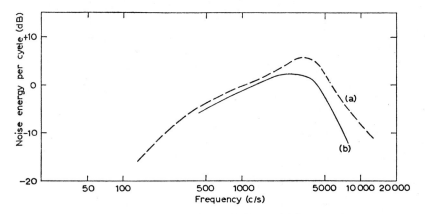

Fig. 7.1.   (a) Spectral response of ear sensitivity to noise.
(b) Spectral response of non-modulated biassed tape weighted with curve (a).
Tape speed $1\frac{7}{8}$ inch/sec.

ing level at high frequencies without incurring excessive intermodulation distortion. Hence the maximum permissible recording level for music can be approximately described as that leading to 5% harmonic distortion at the reproducing head for frequencies below about 3 kc/s, and maximum output at higher frequencies; such characteristics for various coating materials are given in Chapter 6 (Fig. 6.30).

The most noticeable tape noise in recorded music is the background noise of biassed but unmodulated tape. Measurements of noise spectra at the reproducing head, in one third octave band intervals, are also presented in Fig. 6.30. However, the audibility of this background noise will be weighted by the frequency response of the ear sensitivity at a corresponding amplitude level; this is normally taken to be at 40 phon and is plotted in Fig. 7.1 – curve (a) (MCKNIGHT [1959a]). It can be seen that the mid-frequency band is accentuated. A typical reproducing head noise voltage curve from acicular iron oxide tape is replotted in Fig. 7.1 – curve (b), after weighting with the ear sensitivity curve, to give some idea of the subjective spectral response of the un-

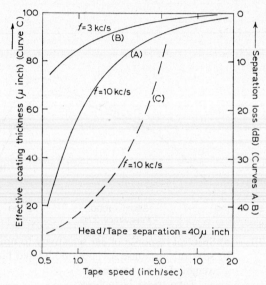

Fig. 7.2.   Dependence of head-tape separation loss and effective recorded layer thickness on tape speed.

equalized reproducing head noise voltage. It is evident that the mid-frequencies will dominate in this noise spectrum resulting in an audible background hiss. The design of an optimum equalization system involves relative apportionment of the frequency responses of the recording and reproducing responses to achieve equal probability at all frequencies of distortion due to overloading in recording, while avoiding accentuation of the mid-frequency noise in reproduction. This topic will be discussed in detail in § 2.3

The overall design specification for a wide-band high-resolution system must weigh the relative advantages of increasing the recorded information density by tape speed reduction or by narrowing the re-corded track width. This is a debatable question especially if probable component advances are anticipated. The limiting factors on speed reduction are the wavelength dependent losses occurring in recording and reproducing, especially those concerned with the separation of the magnetic particles of the coating from the pole-pieces of the recording and reproducing heads. Optimum designs of heads and tapes are aimed at minimizing these losses due to the reduction of recording and repro-ducing resolutions with separation from the heads. The severity of these losses at short wavelengths can be illustrated by considering separation losses in reproduction only, which are indigenous to the system *. In Fig. 7.2, curves A and B, the losses for 10 kc/s and 3 kc/s are plotted respectively as a function of tape speed for a head-to-tape separation of 40 microinches (1 micron). In this case it can be seen that the 10 kc/s out-put drops very rapidly below about 2 inch/sec speed. At 3 kc/s, where the sensitivity to noise is high, the loss is 4 dB at 2 inch/sec.

Some idea of the effective coating thickness contributing to the reproduced signal may be obtained by plotting the equivalent depth of coating corresponding to zero thickness loss. It will be recalled (Ch. 4, § 2.4) that, for a uniformly magnetized tape, the flux in the reproducing head is proportional to

$$\frac{\lambda}{2\pi}\left[1 - \exp\left(\frac{-2\pi c}{\lambda}\right)\right],$$

due to the thickness loss. Providing the wavelength is not large com-pared with the coating thickness, the output is equivalent to that of a coating thickness $(\lambda/2\pi)$ with no thickness loss. Although this equiva-lent thickness is somewhat less than the active layer it gives a guide to the effective coating layer at short wavelengths. The equivalent thick-ness is plotted for 10 kc/s in Fig. 7.2 – curve C. It is evident that the active layer of the tape approaches the assumed surface roughness at

* Wavelength dependent recording losses are also quite significant but are subject to modification by bias control as discussed in § 2.2.

10 kc/s when the tape speed is around 2 inch/sec. Hence a tape speed of 1⅞ inch/sec may presently be considered a possible lower practical limit for audio recording.

Track width reduction also contributes to an increase in the recorded information density. It has been confirmed experimentally that the

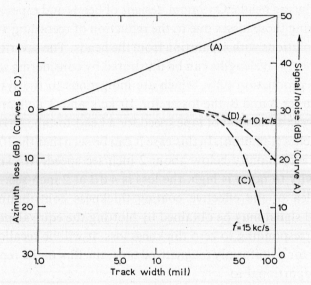

Fig. 7.3.   Signal-to-noise ratio and azimuth alignment losses as function of track width.

theoretical proportionality of the signal-to-noise ratio with the square root of the track width is effective over the width range of 1 mil to 100 mil (ELDRIDGE and BAABA [1961]). Thus, if a 50 dB signal-to-noise ratio is obtained for 100 mil track width, this will reduce to about 30 dB for 1 mil track width, as shown in Fig. 7.3 – curve A. However, there is also an advantage in using narrow tracks in that the azimuth misalignment loss in reproduction (Ch. 4, § 5.2), due to tape skewing during its transit between the recording and reproducing gaps, is reduced. Assuming a skewing angle of 0.0015 radians, the azimuth losses for 10 kc/s and 15 kc/s signals and tape speed of 1⅞ inch/sec vary with track width as shown in curves B and C of Fig. 7.3. A compromise track width of 40 mil is chosen for the system to be described.

§ 2.2.   Optimum Components

Recording Head

The design and operation of the recording head can materially affect the overall signal-to-noise ratio obtainable in a wide-band high-resolution recording system. Long and short wavelength recording processes are considered separately in Chapter 2, § 6; however, the major problem in attempting to record simultaneously long and short wavelengths is linked with the phenomenon of the increase of bias amplitude with wavelength for maximum acceptable output. This is due to the increase of active coating depth with wavelength, necessitating larger ac fields for complete effective magnetization. Short wavelengths, on the other hand, are inefficiently recorded under such conditions since the recording zone, that is, the zone in which the ac bias is in the range which can magnetize the tape irreversibly, is extended, compared with that obtained for optimum short wavelength bias. Such a loss is caused by the reduction of the recording head field decrement, at the trailing pole-piece, with distance from the trailing edge of the recording gap. Severe losses occur when the bias level is large enough to cause the recording zone to extend over a distance comparable with the recorded wavelength.

The optimum design of a recording head, therefore, entails the production of a field contour with a restricted recording zone at the tape surface when the amplitude is sufficient to record through the entire coating thickness. However, it is not necessary to maintain the restricted recording zone for tape layers remote from the head surface, since only the longer wavelengths will be reproduced from these layers. The situation may be best illustrated by considering the shape and extent of the effective recording zones in a similar way to that described for the simple ac bias recording model in Chapter 2. The extent of the recording zone will depend on the dispersion of critical fields required to magnetize the tape irreversibly; in an oriented acicular particle tape this can be assumed to be about 0.25 H_c. The shape of the effective recording zone will depend on the relative sensitivity of the tape to the longitudinal and perpendicular applied field components. It has been shown that, for an oriented tape, the anisotropy of irreversible magnetization is due to both external and internal demagnetizing fields,

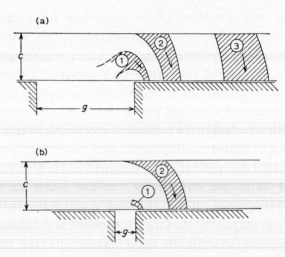

Fig. 7.4.   Recording zones corresponding to total applied field of conventional gapped ring head:
a. Wide gap; b. Narrow gap (ELDRIDGE and DANIEL [1962]).

leading to reduced sensitivity of the perpendicular component. However, it is not possible to completely ignore this component for the predominantly perpendicular fields occurring at the tape surface. Short wavelength recording losses thereby occur, both due to the large amplitude of the bias field causing a spread of the recording zone, and due to the rotation of the field direction in the recording zone towards the perpendicular direction as the tape moves away from the trailing edge of the gap. The effective recording zone therefore depends on the field direction in a rather complicated way. This is approximated in Fig. 7.4*a* by plotting the total equal field contours of the head (ELDRIDGE and DANIEL [1962]), although it is realized that the tape anisotropy will produce some preferential weighting of the longitudinal field component (see Ch. 2, § 2). The shaded zones illustrate three possible recording zones corresponding to different amplitudes of applied field; here the gap length is equal to twice the coating thickness. By comparison, the recording zones for a narrow gap recording head are shown in Fig. 7.4*b*. In this case it can be seen that the narrowest recording zone (1) is achieved, but when sufficient field is applied to magnetize the tape fully, the recording zone is just as large as that from the wide gap head. Hence no advantage is to be gained by use of extremely narrow gaps for wide-band recording. In practice, a compromise bias is chosen which favours the short wavelength response, and gives maximum signal-to-noise ratio in the 3–4 kc/s region where noise is predominantly audible.

Conventional recording head design will employ a moderately narrow gap length to optimize the recording resolution of short wavelengths. It is, of course, just as important to maintain sharp and straight gap edges in the recording head as in the reproducing head to obtain coherence of the recorded magnetization along the width of the reproducing head gap. It is customary to use thin laminations in recording heads, to minimize eddy current losses at the bias frequency, although the concentration of the recording flux in the surface layers of unlaminated poles is not itself detrimental. The bias frequency must be chosen to be high enough to avoid intermodulation with the higher signal frequencies; it must also be high enough to achieve anhysteretic magnetization conditions. Referring to Fig. 7.4 it is at least necessary to ensure that several bias oscillations occur during the transit time of an element of tape across zone (2). For the $1\frac{7}{8}$ inch/sec system considered, a bias frequency of 100 kc/s is sufficient to satisfy this condition.

In considering possible improvements to recording head design a notable advantage would accrue from the suppression of the perpendicular component of the recording field. Not only would the deleterious effect due to rotation of the magnetizing field be removed, but also a high resolution recording zone would be maintained in the tape surface layer where large field amplitudes exist. This is illustrated in Fig. 7.5*a*, where longitudinal field recording zones are plotted, analogous to those for the total field components of Fig. 7.4. It is seen that increasing the applied field produces zones (1), (2) and (3), thus realizing full tape magnetization without extending the recording zone at the tape surface beyond the trailing edge of the gap. Unfortunately, it is very

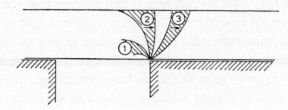

Fig. 7.5a.   Recording zones for longitudinal component of field from a gapped ring head.

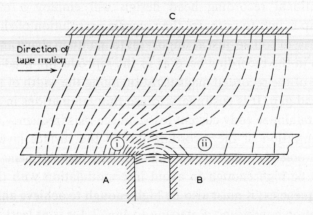

Fig. 7.5b.   Recording field pattern for third-pole head (CAMRAS [1953]).

difficult to achieve complete suppression of the perpendicular field near the gap edge in any practical device. A straightforward method is to apply a perpendicular field from a third pole located on the remote side of the tape coating as illustrated in Fig. 7.5b (CAMRAS [1953]). The direction of the perpendicular field from the leading pole (A) to the third pole (C) aids the perpendicular component of the gap field producing a strong and nearly perpendicular field through the tape in area (i). In area (ii) above the trailing pole-piece the gap field is very nearly annulled by the perpendicular field. Thus the recording zone near the trailing gap edge is made narrower due to a reduction of the perpendicular field component of the gap field. Other designs of recording heads with fields originating from the remote side of the tape have also been proposed which employ gapped pole-pieces on both sides of the tape; some structures use cores to excite both sets of pole-pieces and others rely on induction from excitation of the poles on the remote side of the tape (e.g. CAMRAS [1961], BOGEN and STEINKOPF

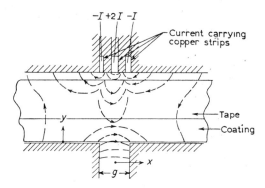

Fig. 7.6a.   Recording head with supplementary quadrupole field source
(GABOR [1962]).

[1960], RÖHLING [1954]). The geometry of such head structures is critical if significant improvements in wide band recording conditions are to be obtained. The aim is to produce a reduction of the decrement of the recording field through the thickness of the coating together with a reduction of the perpendicular field. A further improvement would occur if, in addition, the longitudinal decrement of the recording field could be increased in the tape coating. This has been achieved in a two-head system by the use of a quadrupole field source, rather than the conventional dipole, on the remote side of the tape coating (GABOR [1962]). The quadrupole is achieved with a triple gap pole-piece structure shown in Fig. 7.6a excited with current carrying copper strips placed in the gaps as shown. The centre strip carries a current $+ 2I$

and the outside strips carry $- I$ each. The resulting longitudinal field pattern exhibits reversals at the two outer gap edges as shown on the left side of Fig. 7.6b. However, although these reversals have been reduced to negligible amplitudes in the tape coating, the field decrement is greater than that of a dipole excitation. In fact, at a distance of

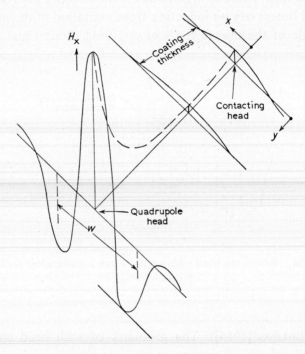

Fig. 7.6b.   Longitudinal recording field distribution (GABOR [1962]).

half a gap length into the coating from the contacting recording pole-pieces, the longitudinal field decrement is about twice that due to the conventional head alone. Furthermore, the field decrement through the coating thickness is reduced by the presence of the supplementary quadrupole head source, as illustrated in Fig. 7.6b.

Recording systems with field sources on both sides of the tape have had some success towards producing optimum bias conditions for a wide range of wavelengths. From a practical point of view it would be simpler to use a single sided recording head structure. In this respect it might be expected that a significant modification of the recording field contour could be brought about by deliberately contouring the recording pole structure. Sharp pole-pieces of the type shown in

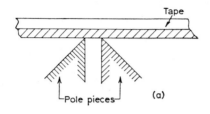

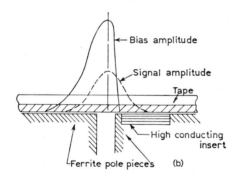

Fig. 7.7.   Modified recording heads for improved field contours:
a. Recording head with sharp poles.
b. Recording head with conducting insert in trailing pole-piece.

Fig. 7.7a do not, however, produce any drastic modification of the field pattern (FRANCIS and KU [1962]). A greater effect may be obtained by replacing the top surface of the trailing pole by a good conductor (WENT and WESTMIJZE [1958]). This acts as an eddy current shield for the high frequency bias and, to a much smaller degree, for the signal. Hence the bias field decrement is greater than that of the signal field, as shown in Fig. 7.7b, tending towards recording conditions approaching ideal anhysteretic magnetization. The increased bias decrement achieved with this can contribute to shorter recording zones and hence to higher recording resolution. With ideal anhysteretic magnetization conditions the decrement of the signal field through the tape coating will cause a corresponding decrease of tape magnetization; in the case of bias and signal falling together, however, the recorded magnetization tends to be uniform through the coating thickness.

Another approach to the problem of providing optimum bias conditions for long and short wavelengths is to employ two conventional recording heads, one following the other, in the same tape path. The first head encountered by a tape element has a relatively wide gap giving a small field decrement through the coating: it is provided with a large bias amplitude and the low frequency end of the signal spectrum is recorded optimally. The second head, having a narrow gap, is provided with a small bias amplitude optimum for recording short wavelengths on the tape surface (WOODWARD [1962], ELDRIDGE and DANIEL [1962]). This second head erases a small part of the long wavelength signal already recorded by the first head. The composite recorded signal may be reproduced by a single reproducing head. Of course, since the two recording heads are spaced along the tape, some time compensation between the two recording signals must be introduced. It would appear that such a recording system can approach the optimum bias for all wavelengths more closely than the composite heads described earlier. However, it should be borne in mind, when assessing any complicated recording head structure, that the advantages of bias optimization for a range of wavelengths diminish as the tape thickness is reduced. Present day trends to thin, high-remanence, coatings may well reduce this problem considerably.

Reproducing Head

The salient features of reproducing head design, described in Chapter 4, § 5, are all applicable to the audio recording system under consideration. As described in Chapter 4, it is necessary to choose pole-piece dimensions which avoid excessive reproduction losses due to the finite length of the pole-pieces relative to the longest recorded wavelength, and due to the gap length between the pole-pieces relative to the shortest recorded wavelength. For a frequency range from 30 to

15 000 c/s the corresponding longest and shortest wavelengths are 60 mil and 0.125 mil, respectively, at a tape speed of $1\frac{7}{8}$ inch/sec. Referring to Figs. 4.17 and 4.8, respectively, it is seen that a total pole-piece length of at least 0.36 inch and a gap length of not more than 0.04 mil (1 micron) are necessary to avoid losses associated with the pole-piece geometry. In practice, the gap length may be increased somewhat, with overall advantage, since the corresponding drop in head coil inductance allows more turns to be used for a given resonant frequency.

The mechanical perfection of the reproducing head gap is, of course, the most critical parameter in minimizing wavelength dependent losses on reproduction. Lapping and polishing techniques have been evolved which make possible the mass production of reproducing heads with effective gaps of 0.04 mil. Although long wavelength losses due to the finite pole-piece length are easily avoided in this system, some care is necessary with the placement of magnetic shields around the reproducing head; these can cause analogous undulations in long wavelength response if they are close to the tape. In the same context, magnetic screens between two ring heads, as might be used to reproduce side-by-side stereophonic audio tracks, must also be carefully designed. In this case it is important that the shield extends up to the pole-piece surface plane, but it must not be so close to the pole-pieces that it shunts some of the desirable core flux. Frequency dependent eddy current losses may easily be reduced to negligible proportions in the reproduction of audio frequencies by using a laminated alloy or ferrite for the core structure.

Thus, in general, it seems that a reproducing head can be designed for efficient transduction of the available tape surface flux. Nevertheless, the losses associated with the effective separation between the reproducing head and the tape and the finite thickness of the coating layer constitute limitations which, as yet, have not been overcome in new reproducing transducer designs using high permeability pole-pieces. Attempts to minimize the separation loss include the use of highly polished pole-pieces; sharply pointed heads have also been used to increase the pressure of the tape at the gap. Other principles for field detection, such as magneto-optical effects, might avoid these limitations but are presently inferior in resolution.

Tape

The general design criteria for magnetic tapes described in Chapter 5, § 1 are applicable to wide-band high-resolution recording systems using ac bias. In addition, the recording properties of various commercial tapes, described in Chapter 6, § 3.3, indicate their relative suitability for recording audio signals at a tape speed of $1\frac{7}{8}$ inch/sec. Little more can be added here except to reiterate the importance of tape design parameters which lead to a reduction of separation losses in recording and reproduction. Any advance in magnetic tape design which allows thinner magnetic coatings to be used, while obtaining the same total flux carrying capacity, will reduce both the short wavelength recording losses due to overbiassing and the thickness losses. Head-to-tape separation losses are minimized by using polished tapes. In addition, accidental separation losses, called dropouts, may also occur due to tape surface blemishes which momentarily lift the tape from the reproducing head. The mechanical perfection of the tape slitting can affect the performance in several ways. Loose shavings from the slitting process can obviously cause separation losses, and non-straight slitting can lead to azimuth misalignment losses and bad reeling properties.

With regard to the magnetic properties of the tape layer, in addition to those discussed in Chapters 5 and 6, it is important to attempt to obtain a reduction in sensitivity to the perpendicular component of the large recording field which occurs at the tape surface in wide-band recording systems. This would lead to improved recording resolution as already described. Existing oriented oxide tapes show some internal anisotropy, but this falls short of the desirable magnitude and is also less than that obtained in some permanent magnet alloys. Further tape developments along these lines should lead to an extension of recording resolution.

§ 2.3. *Equalization Systems*

Even if optimally designed recording and reproducing heads and tape are employed in a wide-band recording system, it is still necessary to apply considerable amplitude correction in order to arrive at a constant transfer characteristic between the audio signal to be recorded, and the reproduced signal. Since, in the end, the signal-to-noise ratio of the system is determined by the signal and noise reproduced from the tape, it is the purpose of the recording equalization to ensure that the tape is magnetized to its maximum permissible value at all frequencies, but especially in the mid-frequency range where the noise is most audible. The equalization subsequently required in the reproducing chain is that necessary to restore equal system gain over the whole frequency range of interest, that is 30–15 000 c/s. The excitation of the reproducing head coil by the core flux, φ_c, may be detected as an open circuit voltage, e, using an amplifier whose input impedance far exceeds the head impedance. Alternatively, it may be read as a short circuit current, i, in which the input impedance is negligible with respect to the head impedance. For each case, respectively,

$$e = n_r (d\varphi_c/dt) \,, \tag{7.1}$$

$$i = n_r \varphi_c / L_r \tag{7.2}$$

where n_r, L_r are, respectively, the reproducing head coil turns and inductance. Although the inherent differentiation of the recording-reproducing system is avoided by reading the short circuited current, the open circuit voltage detection is preferred since better signal-to-amplifier noise ratio is obtained.

The overall unequalized transfer characteristic between the reproducing head voltage, e, and the recording head signal current, i_s, will correspond fundamentally to a 6 dB/octave rise with frequency, modified by the recording and reproducing high frequency losses already described. The relative high and low frequency recording losses depend on the type of recording head used and on the bias amplitude; the latter is normally chosen to favour the mid and high frequency band in order to achieve maximum subjective signal-to-noise ratio. Of course, the predominant losses are those associated with the exponential decrease of the tape surface flux available for detection by the reproducing head.

For the purposes of this development of equalization techniques it will be assumed that the tape magnetization is ideally proportional to the recording head signal current. It will be further assumed that the inherent differentiation in the reproducing head is corrected in the reproducing amplifier. The tape surface flux available for reproduction

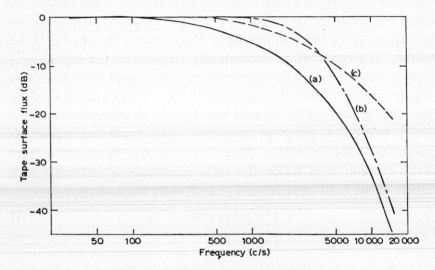

Fig. 7.8.   Tape surface flux *vs.* frequency.
(a) Constant current recording. Recording level – 18 dB on 1 kc/s (5 % distortion).
(b) Maxium tape loading.
(c) Impedance of parallel RC network.
$\tau = 100\mu s$.
Tape speed $1\frac{7}{8}$ inch/sec.

is taken to be the same as the output from a reproducing head whose voltage is integrated and corrected for the relatively small gap and eddy current losses. A typical surface flux characteristic for a $1\frac{7}{8}$ inch/ sec system is shown in Fig. 7.8 – curve (a), corresponding to a constant low-level recording current. The purpose of the equalization system is now to apportion loss correction, corresponding to the departure from constant surface flux in Fig. 7.8, between the recording and reproducing channels to produce maximum overall signal-to-noise ratio and negligible overload distortion. This should be accomplished ideally in such a way that the surface flux spectrum is definable in terms of equivalent electrical circuits. In the case of audio recording it is not necessary to provide phase equalization simultaneously, although circuits have been devised for this purpose for instrumentation recording (BROWN [1960a]).

In order to determine the recording equalization it is initially instructive to compare the spectral response of the tape surface flux for constant current recording, (Fig. 7.8 – curve (a)), with that corresponding to maximum loading of the tape as plotted in curve (b) of Fig. 7.8. For maximum tape loading the previous definition of maximum output is used; this corresponds to 5% harmonic distortion at the reproducing head or to maximum output, whichever is reached first. The tape surface flux curve for optimum tape loading is that which might be obtained typically from a tape consisting of oriented acicular particles of γ Fe$_2$O$_3$. Comparison of curves (a) and (b) of Fig. 7.8 indicates that constant current recording produces less than optimum tape loading in the upper frequency range, and especially so in the mid frequency range. Hence, mid frequency pre-emphasis may be used with corresponding gain in signal-to-noise ratio.

An important factor to be taken into account in the determination of the overall optimum recording equalization is the frequency distribution of the peak energy in the audio signal to be recorded. Any contour in the spectral response of the signal energy may be compensated for in the recording equalization to yield equal probability of overload at all frequencies. It has been shown, however, that this response is highly variable in music and can safely be described only by assuming uniform peak energy at all fequencies in the audio band (McKNIGHT [1959b]). Nevertheless, the probability of equal peak intensities at low and high frequencies is lower than that of the same intensity in the mid frequency band. It has been found by experience that a high frequency pre-emphasis of up to 15 dB at 10 kc/s is tolerable for most music (PIEPLOW [1962]). Special limiting devices may be used to control the rare event of high frequency overload under these conditions.

For the purposes of standardization it is required that the surface flux be describable in terms of the frequency response of a simple

electrical circuit. The response of a parallel combination of resistance
and capacitance having a time constant of 100 μs is plotted in curve
(c) of Fig. 7.8. This particular time constant would appear to be the
most suitable for $1\frac{7}{8}$ inch/sec audio recording, since it corresponds to
about 15 dB of recording pre-emphasis at 10 kc/s. It is to be noted,
however, that additional pre-emphasis of a few decibels in the mid

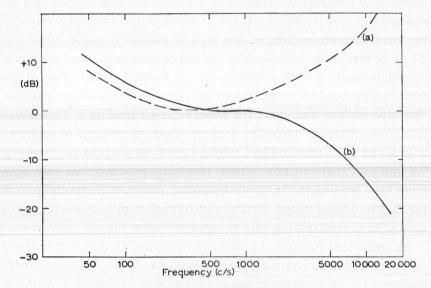

Fig. 7.9.   (a) Recording equalization curve.
(b) Resulting tape surface flux.
Tape speed $1\frac{7}{8}$ inch/sec.

frequency range can be tolerated without overloading (e.g. curves (b)
and (c) of Fig. 7.8). To achieve a further improvement in subjective
signal-to-noise ratio an extra pre-emphasis of several decibels is added
in the range 2–4 kc/s (PIEPLOW [1962], McKNIGHT [1959a], GOLD-
BERG and TORICK [1960]), with a consequent small increase in the probability of mid-frequency overloading. Finally, some low frequency
pre-emphasis is also added to increase the recorded signal level relative
to the power frequency and its harmonics. However, this latter pre-
emphasis is only for the practical purpose of achieving better signal-to-

hum ratio; high quality systems would not benefit from it (MCKNIGHT [1962]). A typical overall recording equalization is plotted in Fig. 7.9, curve (a) along with the corresponding tape surface flux, curve (b).

The equalization required in the reproducing channel includes that necessary to produce constant reproducing head voltage when the tape surface flux spectrum corresponds to that plotted in Fig. 7.9. The re-

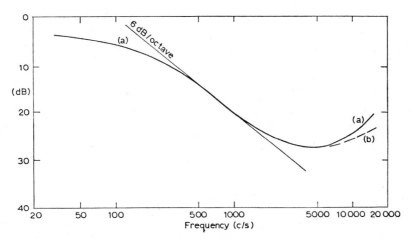

Fig. 7.10.   (a) Reproducing equalization curve.
              (b) Reproducing equalization without gap loss correction.

producing equalization may then be computed from curve (b) Fig. 7.9 together with a 6 dB/octave falling characteristic to compensate for the differentiation in the reproducing head, plus a small correction for the reproducing head gap loss. The computed reproducing channel equalization is then as shown in Fig. 7.10. System performance using the derived equalizations is reviewed in the following section.

§ 2.4.   *System Performance*

The improved recording components and equalization techniques outlined in this chapter have led to the development of practical systems for audio recording at a tape speed of $1\frac{7}{8}$ inch/sec (GOLDMARK *et al.* [1960], PIEPLOW [1962]). For the former system an overall signal-to-noise ratio of 54 dB has been obtained. This refers to a low frequency signal level yielding 3 % harmonic distortion after equalization which is equivalent to 5 % distortion at the reproducing head and is thus representative of the maximum recording level; the associated noise level is that obtained after equalization, ignoring frequencies below 250 c/s. At a recording level of −18 dB relative to the maximum level the response is flat to within 3 dB, from 30–15 000 c/s. Amplitude linearity is obtained up to −6 dB relative to the maximum level for frequencies up to 5 kc/s. At 10 kc/s linearity is obtained up to −18 dB, the maximum output level being −12 dB relative to that at low frequencies. Such a system can record and reproduce most music with little overloading at a sufficiently high level to render the background noise negligible. Of course when the peak energy distribution in the music does not fall off at high frequencies, then compression occurs unless the recording level is reduced. Under these circumstances the noise level will be more apparent. There is room for further developments in tape which will yield lower noise levels when biassed without signal modulation. This would allow recording levels to be lowered, thus increasing the overload capability.

The present rate of progress in the development of the components used in magnetic recording can be expected to be maintained in the future, leading to further advances in high-resolution wide-band recording systems. Improved recording and reproducing methods may also be expected which will extend the use of such systems into many other frequency ranges.

REFERENCES

Agfa, 1954, Brit. Pat., 717, 269.
Aharoni, A., 1959, J. Appl. Phys. **30** 70S.
Amar, H., 1958, Phys. Rev. **111** 149.
Bate, G., 1961, J. Appl. Phys. **32** 239S.
Bate, G., 1962, J. Appl. Phys. **33** 2263.
Bauer, B. B. and C. D. Mee, 1961, I.R.E. Trans. Audio **AU 9** 139.
Bean, C. P., 1955, J. Appl. Phys. **26** 1381.
Beer, H. B. and G. V. Planer, 1958, Brit. Commun. and Electronics **5** 939.
Berkowitz, A. E. and W. J. Schuele, 1959, J. Appl. Phys. **30** 134S.
Blois, M. S., 1955, J. Appl. Phys. **26** 975.
Bockemuehl, R. R. and P. W. Wood, 1960, Electronics **33** 70.
Bogen, W. and W. Steinkopf, 1960, U. S. Pat. 2, 932, 697.
Bonn, T. H. and D. C. Wendell, 1953, U. S. Pat. 2, 644, 787.
Booth, A. D., 1952, Brit. J. Appl. Phys. **3** 307.
Bozorth, R. M., 1956, J. Phys. Radium **17** 256.
Brockman, F. G. and W. G. Steneck, 1954, Philips Tech. Rev. **16** 79.
Brophy, J. J., 1960, I.R.E. Trans. Audio, **AU 8** 58.
Brown, R. F., 1960a, Nat. Bureau Standards Rept. 7023.
Brown, W. F., 1956a, 3M Internal Report.
Brown, W. F., 1956b, 3M Internal Report.
Brown, W. F., 1957, Phys. Rev. **105** 1479.
Brown, W. F., 1959a, J. Appl. Phys. **30** 62S.
Brown, W. F., 1959b, J. Appl. Phys. **30** 130S.
Brown, W. F., 1960b, Amer. J. Phys. **28** 542.
Brown, W. F., 1962, J. Appl. Phys. **33** 1308.
Campbell, R. B., 1957, Boston Conf. Proc. 128.
Campbell, R. B., H. Amar, A. E. Berkowitz and P. J. Flanders, 1957, Boston Conf.
  Proc. 118.
Camras, M., 1949, Proc. Inst. Radio Engrs. **37** 569.
Camras, M., 1953, U. S. Pat. 2, 628, 285.
Camras, M., 1954, U. S. Pat. 2, 694, 656.
Camras, M., 1961, U. S. Pat. 2, 987, 583.
Carman, E. H., 1959, Powder Metall. Bull. **4** 1.
Casimir, H. B. G. *et al.*, 1959, J. Phys. Radium **20** 360.
Cauer, W., 1925, Arch. Elektrotech. **15** 308.
Chapman, D. W., 1963, Proc. I.E.E.E. **51** 247.
Cioffi, P. P., 1950, Rev. Sci. Instrum. **21** 624.
Comerci, F., 1962, I.R.E. Trans Audio **AU 10** 64
Craik, D. J. and E. D. Isaac, 1960, Proc. Phys. Soc. **76** 160.

Craik, D. J. and R. S. Tebble, 1961, Rep. Progr. Phys. **24** 116.

Cummings, R. A., 1962, Private Communication.

Cummings, R. A. and C. D. Mee, 1963. To be published.

Daniel, E. D., 1953, Proc. Instn. Elect. Engrs. **100** 168.

Daniel, E. D., 1960, Private Communication.

Daniel, E. D. and P. E. Axon, 1950, B.B.C. Quarterly **5** 241.

Daniel, E. D. and P. E. Axon, 1953, Proc. Instn. Electr. Engrs. **100** 157.

Daniel, E. D. and P. E. Axon, 1955, B.B.C. Mono. Engng. Div. **2** 4.

Daniel, E. D., P. E. Axon and W. T. Frost, 1957, Proc. Instn. Elect. Engrs. **104B** 158.

Daniel, E. D. and I. Levine, 1960a, J. Acoust. Soc. Amer. **32** 1.

Daniel, E. D. and I. Levine, 1960b, J. Acoust. Soc. Amer. **32** 258.

Daniel, E. D. and E. P. Wohlfarth, 1962, J. Phys. Soc. Japan **17** 670.

Daniels, H. L., 1952, Electronics **25** 116.

Davies, G. L., 1961, Magnetic Tape Instrumentation (New York, McGraw-Hill Book Company) p. 25.

Duinker, S., 1957, Tijdschr. Ned. Radiogenoot. **22** 29.

Duinker, S., 1960, Philips Res. Rep. **15** 342.

Duinker, S., 1961, Philips Res. Rep. **16** 307.

Edmundson, D. E., 1955, Proc. Instn. Electr. Engrs. **102B** 427.

Elder, T. and W. Barte, 1962, Rev. Sci. Instrum. **33** 1360

Eldridge, D. F., 1960, I.R.E. Trans. Audio **AU 8** 42.

Eldridge, D. F., 1961, I.R.E. Trans. Audio **AU 9** 155.

Eldridge, D. F. and A. Baaba, 1961, I.R.E. Trans. Audio **AU 9** 10.

Eldridge, D. F. and E. D. Daniel, 1962, I.R.E. Trans. Audio **AU 10** 72.

Elmore, W. C., 1938, Phys. Rev. **54** 1092.

Fan, G. J., 1960a, Thesis, Stanford University.

Fan, G. J., 1960b, J. Appl. Phys. **31** 402S.

Fan, G. J., 1961a, I.B.M. J. Res. Developm. **5** 321.

Fan, G. J., 1961b, I.B.M. Technical Report No. 17-050.

Feick, G. and H. F. Stedman, 1960, U.S. Pat. 2, 936, 286.

Fisher, R. D. and W. H. Chilton, 1962, J. Electrochem. Soc. **109** 485.

Flanders, P. J., 1957, Boston Conf. Proc. 315.

Flanders, P. J. and S. Shtrikman, 1962, J. Appl. Phys. **33** 216.

Foner, S., 1959, Rev. Sci. Instrum. **30** 548.

Fowler, C. A., E. M. Fryer and D. Treves, 1961, J. Appl. Phys. **32** 296S.

Francis, E. E. and T. C. Ku, 1962, I.B.M. J. Res. Developm. **6** 260.

Frei, E. H., S. Shtrikman and D. Treves, 1957, Phys. Rev. **106** 446.

Freundlich, M. M. *et al.*, 1961, Proc. Instn. Radio Engrs. **49** 498.

Frost, W. T., 1960, I.R.E. WESCON Convention Record, Pt. 5, 46.

Gabor, D., 1962, U. S. Pat. 3,064,087

Gabor, D. and B. B. Bauer, 1962, U. S. Pat. 3, 052, 567.

Gans, R., 1932, Ann. Phys. (Leipzig) **15** 28.

Geyger, W. A., 1956, Electronics **29** 167.

Goldberg, A. A. and E. L. Torick, 1960, J. Audio Engng. Soc. **8** 29.

Goldmark, P. C., C. D. Mee, J. D. Goodell and W. P. Guckenburg, 1960, I.R.E. Trans. Audio **AU 8** 161.

Gould, J. E. and M. McCaig, 1954, Proc. Phys. Soc. B. **67** 584.

Greiner, J., 1953, Der Aufzeichnungsvorgang beim Magnettonverfahren mit Wechselstromvormagnetisierung (Berlin, Verlag Technik) p. 13.

Greiner, J., 1955, Nachrichtentech. **5** 295, 351.

Greiner, J., 1956, Nachrichtentech. **6** 63.

Guckenburg, W. and C. D. Mee, 1961, J. Audio Engng. Soc. **9** 107.

Guillaud, C., 1953, Rev. Mod. Phys. **25** 64.

Hobson, P. T., E. S. Chatt and W. P. Osmond, 1948, J. Iron Steel Inst. **159** 145.

Holmes, L. C. and D. L. Clark, 1945, Electronics **18** 126.

Howling, D., 1956, J. Acoust. Soc. Amer. **28** 977.

Ingraham, J. N. and T. J. Swoboda, 1960, U. S. Pat. 2, 923, 683.

Iwasaki, S. and K. Nagai, 1962, Budapest Magnetic Recording Conf. Proc.

Jacobs, I. S. and C. P. Bean, 1955, Phys. Rev. **100** 1060.

Jacobs, I. S. and P. E. Lawrence, 1958, Rev. Sci. Instrum. **29** 713.

Jacobs, I. S. and F. E. Luborsky, 1957, J. Appl. Phys. **28** 467.

Jeschke, J. C. 1954, E. Ger. Pat. 8684.

Johnson, C. E. and W. F. Brown, 1958, J. Appl. Phys. **29** 1699.

Johnson, C. E. and W. F. Brown, 1961, J. Appl. Phys. **32** 243S.

Jonker, G. H., H. P. J. Wijn and P. B. Braun, 1957, Proc. Instn. Elect. Engrs. **104B** 249.

Karlqvist, O., 1954, K. Tekn. Högsk. Handl. **86** 1.

Kojima, H., 1956, Sci. Rep. Res. Insts. Tohoku Univ. **8** 540.

Kojima, H., 1958, Sci. Rep. Res. Insts. Tohoku Univ. **10** 175.

Koretzky, H., 1963a, First Australia Conf. on Electrochemistry Proc.

Koretzky, H., 1963b, Private Communication.

Kornei, O., 1956, I.R.E. Nat. Convention Record, **4** 145.

Kostyshyn, B., 1962, I.R.E. Trans. Elec. Comp. **EC-11** 253.

Kramer, A., 1956, J. Audio Engng. Soc. **4** 41.

Krones, F., 1960, Technik der Magnetspeicher (Berlin, Springer-Verlag, p. 425.

Lentz, T. and J. J. Miyata, 1961, Electronics **34** 36.

Logie, H. J., 1953, S. African J. Sci. **50** 15.

Luborsky, F. E., 1961, J. Appl. Phys. **32** 171S.

Luborsky, F. E., E. F. Fullam and D. S. Hallgren, 1958, J. Appl. Phys. **29** 989.

Luborsky, F. E., L. I. Mendelsohn and T. O. Paine, 1957, J. Appl. Phys. **28** 344.

Luborsky, F. E. and T. O. Paine, 1960, J. Appl. Phys. **31** 66S.

Luborsky, F. E., T. O. Paine and L. I. Mendelsohn, 1959, Powder Metall. Bull. **4** 57.

Mankin, A. H., 1952, I.R.E. Trans. Audio **AU 9** 16.

Mathes, R. C., 1954, U. S. Pat. 2, 691, 072.

Mayer, L., 1958, J. Appl. Phys. **29** 1454.

McKnight, J. G., 1959a, J. Audio Engng. Soc. **7** 5.

McKnight, J. G., 1959b, J. Audio Engng. Soc. **7** 65.

McKnight, J. G., 1962, J. Audio Engng. Soc. **10** 106.

Mee, C. D., 1950, Proc. Phys. Soc. **63** 922.

Mee, C. D., 1958, Proc. Instn. Electr. Engrs. **105B** 373.

Mee, C. D., 1962, I.R.E. Trans. Audio, **AU 10** 161.

Mee, C. D. and J. C. Jeschke, 1963, J. Appl. Phys. **34** 1271.

Meiklejohn, W. H., 1953, Rev. Mod. Phys. **25** 302.

Miyata, J. J. and R. R. Hartel, 1959, I.R.E. Trans. Elec. Comp. E.C. **8** 159.

Mones, A. H. and E. Banks, 1958, J. Phys. Chem. Solids, **4** 217.

Morrish, A. H. and L. A. K. Watt, 1957, Phys. Rev. **105** 1476.

Morrish, A. H. and S. P. Yu, 1956, Phys. Rev. **102** 670.

Nagai, K., S. Iwasaki and T. Moriya, 1959, Joint Meeting of Elec. Inst. No. 734.

Nagai, K., S. Iwasaki and T. Moriya, 1960a, Private Communication No. 3-1-13.

Nagai, K., S. Iwasaki and T. Moriya, 1960b, Private Communication No. 449.

Néel, L., 1943, Cahiers de Phys., **17** 47.

Néel, L., 1947, C. R. Acad. Sci. (Paris) **224** 1550.

Néel, L., 1949, Ann. Géophys. **5** 99.

Néel, L., 1955, Phil. Mag. **4** 191.

Néel, L., 1958, C. R. Acad. Sci. (Paris) **246** 2313.

Néel, L., 1959, J. Phys. Radium **20** 215.

Oppegard, A. L., F. J. Darnell and H. C. Miller, 1961, J. Appl. Phys. **32** 184S.

Osmond, W. P., 1952, Proc. Phys. Soc. B. **65** 121.

Osmond, W. P., 1953, Proc. Phys. Soc. B. **66** 265.

Osmond, W. P., 1954, Proc. Phys. Soc. B. **67** 875.

Pauthenet, R., 1950, C. R. Acad. Sci. (Paris) **230** 1842.

Pearson, R. T., 1961, Proc. Instn. Radio Engrs. **49** 164.

Persoon, A. H., 1957, Sound Talk Bull., No. 35.

Pieplow, H., 1962, I.R.E. Trans. Audio **AU10** 34.

Preisach, F., 1935, Z. Phys. **94** 277.

Price, R. L., 1958, I.R.E. Trans. Audio **AU6** 29.

Rabe, E., 1960, Brit. Pat. 830, 736.

Rassmann, G. and O. Henkel, 1961, Phys. Status Solidi (E. Germany) **1** 517.

Rimbert, F., 1957, C. R. Acad. Sci. (Paris) **245** 406.

Röhling, H., 1954, U. S. Pat. 2, 675, 429.

Sallo, J. S. and J. M. Carr, 1962, J. Appl. Phys. **33** 1316.

Sallo, J. S. and K. H. Olsen, 1961, J. Appl. Phys. **32** 203S.

Schmidbauer, O., 1957, Elektron. Rdsch. **11** 302.

Schmidt, E., 1960, J. Audio Endgng. Soc. **8** 52.

Schools, R. S., 1961, J. Appl. Phys. **32** 42S.

Schuele, W. J., 1959, J. Phys. Chem. **63** 83.

Schuele, W. J. and V. D. Deetscreek, 1961, J. Appl. Phys. **32** 235S.

Schwantke, G., 1957, Acustica **7** 363.

Schwantke, G., 1961, J. Audio Engng. Soc. **9** 37.

Shtrikman, S. and D. Treves, 1959, J. Phys. Radium **20** 286.

Shtrikman, S. and D. Treves, 1960, J. Appl. Phys. **31** 58S.

Slonczewski, J. C., 1958, J. Appl. Phys. **29** 448.

Smaller, P., 1959, J. Audio Engng. Soc. **7** 196.

Smit, J., H. P. J. Wijn, 1959, Ferrites (New York, J. Wiley) p. 157, 194, 208.

Spratt, H. G. M., 1958, Magnetic Tape Recording (New York, Macmillan) p. 140.

Stanley, C. B., 1960, I.R.E. Trans., Sp. Elec. & Tel. **6** 19.

Stein, I., 1961, I.R.E. Trans. Audio **AU9** 146.

Stein, I., 1962, I.R.E. Nat. Convention Record **7** 42.

Stewart, W. E., 1958, Magnetic Recording Techniques (New York, McGraw-Hill Book Co.).

Stoner, E. C. and E. P. Wohlfarth, 1948, Phil Trans. Roy. Soc. **240** 599.

Stuijts, A. L. and H. P. J. Wijn, 1958, Philips Tech. Rev. **19** 209.

Sugiura, Y., 1960, J. Phys. Soc. Japan **15** 1461.

Swoboda, T. J. et al., 1961, J. Appl. Phys. **32** 374S.

Teer, K., 1961, Philips Res. Rep. **16** 469.

Tonge, D. G. and E. P. Wohlfarth, 1958, Phil. Mag. **3** 536.

Torkar, K. and O. Fredriksen, 1959, Powder Metall. Bull. **4** 105.

Ugine, 1947, Brit. Pat. 590, 392.

Ugine, 1948, Brit. Pat. 596, 875.

Underhill, E. M., 1948, Electronics **21** 122.

Van Oosterhout, G. W., 1956, Appl. Sci. Res. B. **6** 101.

Van Oosterhout, G. W., 1960, Acta Cryst. **13** 932.

Van Oosterhout, G. W. and C. J. M. Rooijmans, 1958, Nature **181** 44.

Von Behren, R. A., 1955, J. Audio Engng. Soc. **3** 210.

Von Behren, R. A., 1956, Sound Talk Bull., No. 34.

Von Behren, R. A., 1958, Sound Talk Bull., No. 37.

Wallace, R. L., 1951, Bell Syst. Tech. J. **30** 1145.

Watt, L. A. K. and A. H. Morrish, 1960, J. Appl. Phys. **31** 71S.

Weiss, P. and R. Forrer, 1929, Ann. Phys. (Paris) **12** 297.

Went, J. J. and W. K. Westmijze, 1958, U. S. Pat. 2, 854, 524.

Werner, P. H. and E. Kohler, 1960, Tech. Mitt. P.T.T. **38** 217.

Westmijze, W. K., 1953a, Philips Res. Rep. **8** 161.

Westmijze, W. K., 1953b, Philips Res. Rep. **8** 245.

Westmijze, W. K., 1953c, Philips Res. Rep. **8** 343.

Wiegand, D. E. and W. W. Hansen, 1947, Trans. Amer. Inst. Electr. Engrs. **66** 119.

Winckel, F., 1960, Technik der Magnetspeicher (Berlin, Springer-Verlag).

Wohlfarth, E. P., 1955, Proc. Roy. Soc. A **232** 208.

Wohlfarth, E. P., 1956, J. Roy, Coll. Sci. **26** 19.

Wohlfarth, E. P., 1958, J. Appl. Phys. **29** 595.

Wohlfarth, E. P., 1959a, Advances in Phys. **8** 87.

Wohlfarth, E. P., 1959b, J. Appl. Phys. **30** 117S.

Wohlfarth, E. P., 1959c, J. Phys. Radium **20** 295.

Wohlfarth, E. P., 1960, Phil. Mag. **5** 717.

Wohlfarth, E. P. and D. G. Tonge, 1957, Phil. Mag. **2** 1333.

Woodward, J. G., 1962, J. Audio Engng. Soc. **10** 53.

Woodward, J. G. and E. Della Torre, 1959, J. Audio Engng. Soc. **7** 189.

Woodward, J. G. and E. Della Torre, 1961, J. Appl. Phys. **32** 126.

Woodward, J. G. and M. Pradervand, 1961, J. Audio Engng. Soc. **9** 254.

Zenner, R. E., 1951, Proc. Instn. Radio Engrs. **39** 141.

AUTHOR INDEX

SUBJECT INDEX